Routledge Revivals

Modern Materialism and Emergent Evolution

Originally published in 1929, McDougall examines the pertinent conflict between religion and science. His work exhibits the failure of scientists to explain human action mechanistically (the essence of modern materialism), establishes purposive action as a type of event radically different from all mechanistic events, and justifies the belief in teleological causation without which there can be neither religion nor morals. This title will be of interest to students of both the Humanities and Sciences, particularly those studying psychology and philosophy.

Modern Materialism and Emergent Evolution

William McDougall

Routledge
Taylor & Francis Group
ROUTLEDGE REVIVALS

First published in 1929
by Methuen & Co., Ltd.

This edition first published in 2016 by Routledge
2 Park Square, Milton Park, Abingdon, Oxon, OX14 4RN
and by Routledge
711 Third Avenue, New York, NY 10017

Routledge is an imprint of the Taylor & Francis Group, an informa business

Publisher's Note
The publisher has gone to great lengths to ensure the quality of this reprint but points out that some imperfections in the original copies may be apparent.

Disclaimer
The publisher has made every effort to trace copyright holders and welcomes correspondence from those they have been unable to contact.

A Library of Congress record exists under LC control number: 29014453

ISBN 13: 978-1-138-19262-1 (hbk)
ISBN 13: 978-1-315-63979-6 (ebk)

MODERN MATERIALISM AND EMERGENT EVOLUTION

BY

WILLIAM McDOUGALL, M.B., F.R.S.

PROFESSOR OF PSYCHOLOGY IN DUKE UNIVERSITY, FORMERLY
PROFESSOR IN HARVARD COLLEGE AND READER
IN THE UNIVERSITY OF OXFORD

METHUEN & CO. LTD.
36 ESSEX STREET W.C.
LONDON

First Published in 1929

PRINTED IN GREAT BRITAIN

PREFACE

THE question of the reality of teleological action has been remarkably neglected. We have had much discussion of the evidence for design in Nature; but those who incline to regard that evidence as respectable, equally with those who regard it as convincing, have seldom examined the prior and more fundamental question, namely: Is there good reason to believe that causal processes are or can be in any instances governed by design? Among contemporary thinkers, the vast majority of men of science and, I think, a considerable majority of philosophers, assume that all causation is of one type only, namely, the mechanistic type, that teleological causation does not occur, or, at least, that all instances of seemingly teleological causation are but specially complicated and obscure forms of mechanistic causation. The dominance of modern thinking by this assumption is the essence of what in this book is meant by Modern Materialism.

Many men of science are quick to disclaim Materialism; and the exponents of Emergent Evolution, though differing widely in other respects, are pretty well agreed in asserting that this new and fashionable doctrine is not mechanistic. But the repudiation of Materialism generally means merely the repudiation of the now old-fashioned Atomic Materialism of earlier centuries; and the assertion that contemporary evolutionary theory is not mechanistic means merely the acknowledgment of the inadequacy of the theory of causation that was the natural ally of Atomic Materialism, the theory that all causation consists in the imparting of motion or momentum from one body to another.

The denial of the reality of teleological causation is, then, the characteristic feature of contemporary science and justifies the designation Modern Materialism. In these lectures I have sought to show that, when Atomic Materialism is rejected, the remaining grounds for denying teleological causation are very flimsy; that Science is in the very act of repudiating the chief remaining ground, namely, the belief that conscious thinking cannot affect the course of physical events; and that we have the strongest possible grounds for believing that our own voluntary actions are instances of truly teleological intervention in the course of physical events.

My *Body and Mind* was chiefly concerned to establish the causal efficacy of the events that constitute our psychical life. That point seems to be already conceded, or in a fair way to be generally conceded, by men of science. But, in the eighteen years that have elapsed since the writing of that book, I have realized more and more clearly that, in accordance with the prevalent fashion, I had grossly neglected the problem of teleological causation and that the careful examination of this problem must yield very strong, perhaps the strongest of all, arguments in support of the thesis of the book. The present volume is then a supplement to my

Body and Mind; it endeavours to make good the neglected part of the argument. But it is quite capable of standing alone and does not presuppose on the reader's past acquaintance with the earlier volume.

I have not attempted to point to the moral and religious bearings of the conclusion I have sought to establish. They seem to me sufficiently obvious. All talk of religion and morals in a purely mechanistic world seems to me mere *flatus vocis*, in spite of the many philosophers, from Spinoza to Lloyd Morgan, who have tried to persuade us to the contrary opinion.

I may add for the benefit of those who are interested in the great problem of the relation between the physical and the mental, that, if I were now rewriting my *Body and Mind*, I should modify it in the direction indicated in the concluding chapters of my *Outline of Abnormal Psychology* and in my presidential address to the Society for Psychical Research. If in this volume I have said little about Psychical Research, it is not that I hold its value or significance to be slight, but that I wish to establish teleological causation by an argument from indisputable and generally accepted facts of empirical observation.

The limitation imposed by the lecture form has made it impossible to include in the text criticisms of various modern writings relevant to my topic. I have, therefore, appended such criticisms in the form of notes to the several lectures, as well as a few notes that may usefully supplement some highly condensed parts of the text. However, only one of these, the long note on *Some Attempts to Exhibit Teleological Causation as Crypto-mechanistic*, forms an essential part of the argument.

It has been an invidious part of my task to criticize severely a number of writers whom I hold in high esteem and some of whom I am proud to call my friends. That, however, was inevitable and I can but offer them my apologies if in any instances my criticism may seem harsh or unfair.

I have to return cordial thanks to the President, Faculty and Students of the Louisville Presbyterian Seminary for the encouragement given me by their cordial hearing and to the governing body for the appointment as Duncan Lecturer, an appointment which forced me to summon up courage to undertake a very difficult task. I have also to thank the Editor of *Mind* for permission to reprint the note *On the Consonance of Welfare and Pleasure.*

W. McD.

DUKE UNIVERSITY, N.C.
November 1928

These Lectures were delivered at the Louisville Presbyterian Theological Seminary on the William G. Duncan Lectureship in Religious Education, in November, 1928.

CONTENTS

MODERN MATERIALISM AND EMERGENT EVOLUTION

CHAPTER I

INTRODUCTION

IN the second half of the nineteenth century there raged a fierce battle between Science and Religion. In that battle Science seemed at first to win all along the line. The defenders of Religion were forced to concede that the Scriptures may not be interpreted quite literally as they had been by simple pious souls; that perhaps the physical universe had not been created in six days; that perhaps the sun did not even seem to stand still in the heavens upon a certain day, at least not to those who observed its course in calm detachment from the excitement and weariness of battle. It was even conceded by them that, perhaps, the creation of man had been no instantaneous but, rather, a slow and gradual process, a process which might not inaptly be described as one of evolution. In fact, the defenders of Religion were driven to admit that the Holy Scripture, though it might be in some sense an impregnable rock, had outlying parts which could not be effectively defended, but were rather points of weakness to the defenders; that, as with other impregnable rocks, the position would be the more certainly impregnable, if the defenders should withdraw from these indefensible outworks and concentrate their forces upon the central citadel.

In face of many such concessions, the champions of Science were mollified. No longer hard-pressed to maintain their right to exist, to pursue their researches, to announce their conclusions frankly to all the world, they found time to rest, now and then, on their dripping blades and, so resting, to think. And when they thus found time to think, they discovered that all their labours, all their successes in storming this and that position of orthodox religion, had not enabled them to construct a complete, consistent and wholly intelligible account of the nature and genesis of man and of the multitude of humbler living things. Further, about the end of the century, it became clear to the more perspicacious of the men of science that, even as regards the inanimate physical world, the account given by Science of its nature and genesis was very imperfect.

It was not merely that physical science could give no plausible suggestion as to the beginnings of the physical world, that it was still confronted by the antinomies of boundless space and endless time ; it was not only that the nature of Life remained obscure and that Mind, which had constructed the account of the physical universe, was found to remain outside the picture, an irreducible surd ; it was rather that a much more homely and intimate defect of the whole account rendered by physical science began to thrust itself upon the attention of the physical scientists.

THE PASSING OF ATOMIC MATERIALISM

The scheme of things which, towards the middle of the nineteenth century, had become the official creed of Science was one which had been formulated long ago by an ancient Greek, the philosopher Democritus ; had been restated and bathed in the glamour of fancy by the Roman poet Lucretius ; in the seventeenth century had been given new life by the works of the illustrious mathematicians and astronomers, Kepler and Galileo ; and had been developed by a long line of their successors until it had been rounded out with convincing clarity by the great Newton. It was the scheme of Atomic Materialism, mechanical materialism in the strictest sense. According to this scheme the physical world consists, without remainder, of atoms, or minute particles of solid matter, in motion ; all energy is the momentum of such atoms ; and all happenings, all events, all processes, are their motions and accelerations ; and all changes of motion are produced by, all influence, all causation consists in, the impact of one particle upon another.

Atomic Materialism was propounded as an account of the physical universe. It made no claim to describe, to account for, or to abolish, the world of Mind or Spirit. The great Newton himself was a deeply religious man and was much occupied with reflection upon the spiritual world. Descartes, who had played an important part in developing Atomic

Materialism, gave precision to the distinction between Mind and matter, making of them two worlds, disparate and wholly apart, except for their interactions in certain highly peculiar organs, the brains of living, thinking men. Most of their contemporaries agreed with them in regarding the world of Mind as at least as real and active as the world of matter ; while many of them were deists, and saw in the harmonies and marvels of the physical world sure evidence that it had been designed by the mind of a Creator. But, with the continually renewed successes of physical science, the material world of atoms in motion predominated more and more in the minds of scientific men in respect of reality and efficacy. The world of mental reality was more and more regarded as shadowy and relatively unreal. Already in the eighteenth century there were not wanting bold iconoclasts who proclaimed that the material world was alone real ; that the mental world, in so far as it existed, was entirely dependent on the world of atoms, a by-product of its motions. And in the third quarter of the nineteenth century, this view, or something very like it, was accepted by a large number, probably a majority, of worthies in both the physical and the biological sciences.

Few men of science, perhaps, of that date would have subscribed to the eighteenth-century dictum that 'the brain secretes thought as the liver secretes bile'. But the belief in the truth of Atomic Materialism was a principal factor in leading to the denial of all influence of Mind upon the physical world, including the parts and processes of the human body. We see that fact clearly illustrated by the words of a great physicist of that date (John Tyndall); when, in arguing against all influences of Mind on body, he asked: 'Can we imagine two particles in the brain bound together by the thought of a beefsteak?'

This virtual banishment of Mind from the world, or at best the reduction of it to a mere spectator, if not a mere appendage and by-product of material events, was a consequence of Atomic Materialism sufficiently startling to give pause to all those of its exponents who were not biased by positive hostility to the Church. But this consequence would not have sufficed to force a revision of the scheme. If Atomic Materialism had proved itself the best, or the only satisfactory, scheme of the physical world, a scheme self-consistent, complete and intelligible, it would have continued to flourish despite all consequences, and to force itself upon the acceptance of all educated men.

As was said above, it was the discovery of a homely and intimate defect of this grandiose scheme, rather than any bizarre and disturbing deductions from it, that checked its triumphant march. With further advance of the physical sciences, it became obvious that, merely as a scheme descriptive of the physical word, Atomic Materialism was wofully incomplete; that it omitted to mention, and offered no explanation of, and no prospect of explanation of, a number of very important facts which the physical sciences had themselves revealed—the facts of gravitation, of magnetic attraction, of chemical affinity and synthesis, of latent energy, of what on the face of it seemed to be action at a distance, manifested in many phenomena.

The physicists of the later nineteenth century were largely occupied in endeavours to patch the scheme of Atomic Materialism, by adding supplementary hypotheses. One new feature after another was added ; the chief being the universal ether, a sort of immaterial and non-atomic matter. By postulating for this ether various constitutions, implying various impossible conjunctions of properties, such as perfect rigidity and elasticity, absence of mass and gravity and friction, it was sought to make the scheme of Atomic Materialism workable, to make the garment waterproof. But, when the patching had gone on for some time, it became obvious that but little of the original cloth remained ; the scheme began to appear no longer as a faithful photographic representation of the physical world ; it took on rather the aspect of an ingenious system of man-made working hypotheses, useful for the guidance of research, but containing so many obvious gaps and inconsistencies, suggesting forcibly so many unanswerable questions, that it could no longer be seriously presented as either a faithful description or as a comprehensive system of explanation of all physical events.

The physicists began to see, not only that physical events could not be adequately described or explained in terms of matter and its motions in space, but also that the addition or the substitution of various energies, active and potential, and of the mysterious ether, did not suffice to restore to Atomic Materialism its lost respectability. They realized that neither matter in motion, nor energy, nor electrons, nor ether, stressed, strained or twisted or otherwise tortured, were the only sufficient and therefore indispensable ultimates or indefinables, the postulation of which could render useful service to physical research. The simple indivisible atom of Democritus has become a fantastic and vastly complex whirl of entities or events, the description of which undergoes radical changes every few years. Compared with the fashions in atomic constitution the fashions in ladies' headgear seem stable and familiar.

THE TRUCE BETWEEN SCIENCE AND RELIGION

Atomic or strictly mechanical Materialism has, then, been undermined and swept away by the progress of the physical sciences themselves. Now, men of science are not a species apart; they are not necessarily oblivious of the moral and religious needs of mankind. Hence, when the dissolution of Atomic Materialism had become patent, when the physicists had realized that the progress of physical science raised at least as many new problems as it solved, problems which were capable of merely provisional solutions in terms of many widely different hypotheses, they were glad to call a truce to the ancient warfare against Religion. The practical successes of their work had assured them the right of free inquiry and free speech; and they were prepared to seek a peace without victory. Hence, of recent years we have heard much smooth speech of the reconciliation of Science with Religion; we are told that henceforth they may walk hand-in-hand without risk of discord. We see the lion and the lamb lying down together. For Science, which in the time of Galileo cowered beneath the roars of the Church, has become the lion ; while the Church has become the shrinking lamb, anxiously watching its steps lest it provoke its mighty companion. The lion, his appetite sated by many a full meal, looks benevolently upon the lamb. At the present day the venerable statesman who naïvely defended the Bible miracles against the ferocious attacks of T. H. Huxley is assimilated in memory with Christians in the Colosseum of Nero ; and even the late W. J. Bryan's defiance of Science provokes in the modern breast that sympathetic respect which is always due to a good man struggling against great odds.

But though Science has turned pacifist and though Religion is prepared for defensive warfare only, the conflict is not yet ended. Materialism of the cruder sort is no longer respectable ; but its place has been taken by a doctrine or a point of view, which, though it does not so flatly negate all religious hopes and moral aspirations, cannot easily be reconciled with either. Materialism in the literal sense has gone, never to return ; but Science still renders an account of Man and the universe which, if not positively hostile, is yet adverse to every form of Religion, however broadly defined, and obstructive to every form of moral effort. And this account is the more dangerous to these interests just because it avoids the crudities of the older Materialism and is rendered in subtle and elusive terms, in terms difficult to grasp or define, and therefore in propositions difficult to refute.

MODERN MATERIALISM

In the course of the nineteenth century, Atomic Materialism had invaded the young science of biology, permeated and shaped its teachings, and received from it new confidence. Two great fields of biological research yielded results that seemed to harmonize with it, and therefore to confirm it. Physiology claimed to show that the functioning of all the organs of the human body, especially of the brain, could be explained mechanically, in terms consistent with Atomic Materialism ; it seemed thus to render untenable Descartes' theory that the human brain was the seat of interaction between the world of matter and the world of Mind, that in and through the brains of living men intelligent purpose could and did effectively intervene in the course of physical events, bending them toward the fulfilment of men's desires. Secondly, Darwin's theory of the origin of species by natural selection led to the general acceptance of the theory of organic evolution and made it seem that all the marvels of nicely adapted structure and function displayed by living things were the products of a purely mechanical process of adaptation continued through many millions of years.

Thus Atomic Materialism engendered a Biological Materialism that harmonized with it and seemed to confirm it. Atomic Materialism was in origin and substance a theory of the inorganic world. Biological Materialism was an extension of the same general type of explanation to the sphere of living things, to man himself, his every thought and action. And, now that Atomic Materialism has collapsed in ruins, Biological Materialism remains, confident of its own findings and still claiming the support of the physical sciences.

Though physical science has abandoned Atomic Materialism, it has not hitherto revealed in the events of the inorganic world any clear indication of mental life, any trace of intelligent purposive activity. The physical sciences are no longer mechanical ; but they are still mechanistic in the wider sense of that word.[1] For we may properly apply the term mechanistic to every explanation or description of the course of events that is not teleological, to every explanation or description that traces the sequence of events from past to present, leaving aside all reference to future possibilities in accounting for present process. In this sense, then, the physical sciences, though no longer mechanical, are still mechanistic ; and biology, as commonly practised, basing itself on the physical sciences, attempts to explain the processes of living things by speaking and thinking about them in the language of those sciences. And even of those who concern themselves with the processes of the human mind a large proportion persists in the attempt to describe, or even to explain, them in the same terms.[2]

[1] Note I, *On the Meaning of 'Mechanistic'*.

[2] The biologists who follow this practice, inevitably find difficulty in keeping up with the rapid changes of contemporary physical and chemical science ; hence their language and ways of thinking are apt to represent, not strictly contemporary physics and chemistry, but those of a bygone period ; and, as a

Such, then, is the nature of Modern Materialism. It consists essentially in the attempt to describe and explain all processes, including those of Life and Mind, in terms of a physical science that knows nothing of Life and Mind, that has deliberately for its own purposes abstracted from, or refused to consider and take account of, the facts of Life and Mind. We may, I suggest, properly continue to designate as Materialism any and every attempt to extend mechanistic descriptions or explanations to all events, i.e. descriptions and explanations that (in accordance with the negative definition of ' mechanistic ', which we have seen to be the only comprehensive one) are non-teleological, that take no account of processes of the type with which we are most immediately and familiarly acquainted, namely, our own mental processes. For such a mechanistic scheme of things in general is the lineal descendant and true successor of Atomic Materialism; and, more importantly, it carries the same practical consequences, points to the same general conclusions, namely, a view of the world and all its processes from which Mind as an active factor, a real agent or activity that makes a difference to the course of events, is wholly excluded, save perhaps as a shadowy spectator on the side lines, a spectator whose applause or other comments on the game are not even heard by the players, or, if heard, are wholly unheeded.

Mechanistic science, then, is science that excludes, neglects, ignores or abstracts from the processes of

matter of fact, many of them are still thinking implicitly in the manner proper to atomic materialism.

Mind; and Modern Materialism is the assumption that such mechanistic science can in principle achieve a complete and satisfactory account of the world and of man, his nature, origin and destiny. The assumption is widely accepted both by men of science and by philosophers; and the question whether it is well founded is the most important and burning question that confronts the mind of man at the present time.

ITS PRACTICAL CONSEQUENCES

I said just now that Modern Materialism and Atomic Materialism lead to the same practical consequences. Let us pause a moment to consider more nearly what those practical consequences are. First, let us note that, if the initial assumption of Modern Materialism is false as regards both organic and inorganic nature, then all the sciences, in so far as they are mechanistic, are on a false scent and need to be radically reformed; and if it is true of the inorganic or physical realm and untrue of the organic, then the physical sciences may progress indefinitely on their present lines, while the biological sciences need be reformed if they are to make further progress.

Secondly, our view of the nature, origin, role and destiny of mankind is profoundly affected. *If the mechanistic assumption is valid,* we cannot validly postulate any, even the slightest, degree of freedom of choice, or any effectiveness of our ideals and of our aspirations for their fulfilment; we cannot believe in the reality of moral effort or of creative activity of any kind; our belief that we can by our efforts contribute in some degree to the realization of our ideals; our belief that by taking thought we may refine our ideals, or give preference to the better over the less good; our belief that by self-discipline and culture we may raise ourselves in some degree in the scale of personal value and contribute however little towards the conservation of values—all such beliefs are illusory.

Thirdly, the mechanistic assumption is incompatible with all those religious beliefs which to the plain man are of the very essence of religion, those beliefs the destruction of which would leave his religion (if it at all survived) merely a system of pale abstract propositions, propositions incapable of evoking enthusiasm, devotion, loyalty or reverence, however sincerely he might believe them. I do not assert that Modern Materialism is incompatible with and necessarily destructive of all religions. Like Atomic Materialism it may logically comport with Pantheism, whether in the form adopted by Spinoza or in the Absolutist form of the Hegelians; or with Deism, the theory of a great Designer, who, with infinite skill and knowledge, has created the universe as one vast machine of which Man is but one insignificant part.

But let us notice that, though the mechanistic assumption is logically compatible with Deism, it destroys the principal foundation of that theory. For the deistic theory regards the Deity as a designing Creative Mind. Now, since the only minds with which we are nearly acquainted are our own, we can conceive such a Deity only after the pattern of our own minds, only in the light of our own experience of creative mental activity. If, then, with Modern Materialism, we regard all belief in such activity as illusory, we have no positive and sufficient ground for believing that any such activity is real or possible.[1]

The issue raised by the prevalence in the scientific world of Modern Materialism is, then, not only a most difficult theoretical puzzle, fascinating to speculative minds by reason of its difficulty; it is also of the greatest practical importance as affecting most intimately the conduct of our lives, our governing beliefs, our hopes, our fears, our aspirations, and our efforts.

THE ISSUE LIES WITH SCIENCE, NOT WITH PHILOSOPHY

To what discipline shall we turn for the solution of this problem? From what class of students may we rightly demand guidance in our endeavour to cope with it? Where lies the responsibility? What branch of learning should formulate the alternative

[1] I cite in this connexion the apposite remarks of a stalwart mechanist, Professor H. C. Warren: 'The philosopher will obtain scant hearing among scientists if, after conceding the mechanistic character of biological events, he attempts to smuggle entelechy [i.e. some purposive directive power] in at the beginning of things. A notion accepted in one sphere may be carried over by analogy into another with some plausibility. But if it afterward prove false in the former sphere, one should demand independent evidence if it is to stand in the latter. If the action of a directing agency during the *course* of events is unsupported by evidence, there is no *a priori* ground for assuming such a directive agency at the *beginning* of events.'—'A Study of Purpose,' *Journal of Philosophy*, vol. XIII.

possibilities and impartially guide us to the choice of the best, the most valid, the best founded, the most consistent with all relevant facts ?

Clearly, it is of little use to turn to the physical scientists. A Newton, a Pupin, a Lodge, may tell us impressively of his religious and moral convictions ; but these convictions are not the conclusions to which he is led by his physical researches ; they concern a sphere of which his branch of science has no knowledge, a sphere in which his lifelong studies in another sphere bring him no special competence, his profound erudition and eminence no authority.

You will say : Obviously we must turn to Philosophy and the philosophers. I answer : If you so turn you will be disappointed. Philosophy has no authoritative answer to give. You will find among modern philosophers the greatest possible diversity of views. It is, I think, roughly true to say that, in the age of Atomic Materialism, about half of the then flourishing philosophers accepted it ; that at the present day about the same proportion accepts the mechanistic assumption ; and that, among the other half, the majority consistently hedge on this great question. I do not make this a reproach against them. I think they are right to hedge, that is to say, to keep their minds open and their systems so plastic as to be capable of reconciliation with either alternative. For I would maintain that the decision of this question does not lie within the province of Philosophy rightly understood.

The question at issue is a question of fact, namely : Does, or does not, Mind play a real part in the world ? Is the seemingly teleological activity of our minds a real and effective factor ? And, if it is a real and effective factor in the course of events, is it really teleological ? Or is it merely a form of mechanistic process in disguise, one that falsely appears to us to be an effective striving for the realization or the attainment of desired goals ? Philosophy is properly concerned, not with questions of fact, but with questions of value. The determination and discovery of facts, the building of knowledge, is the province of Science. Philosophy is directly concerned, not with the construction of knowledge, but with the formation and refinement of opinion ; not with the ascertainment of what is, but with the determination of what ought to be : for values are and must ever remain matters, not of knowledge, but of opinion. To the question : What ought to be or to be done ? Philosophy has for answer no categorical imperative. Only revealed Religion can return such answers.

It is true that in the past those who were popularly known as philosophers did seek to pronounce on matters of fact ; as when the early Ionian philosophers pronounced earth, air, fire and water to be the elements of all things. And, so long as the spheres of Science and Philosophy were not clearly defined, the study of physical science was known as natural philosophy. It is true also that, to the present day, philosophers are very apt to attempt to deduce answers to questions of fact from their opinions upon relative values ; and, perhaps, where knowledge is not attainable by Science such deduction has a certain validity ; that is the right claimed by William James in his justly celebrated essay on *The*

Will to Believe. But such pronouncements of the Will to Believe must be held as provisional only ; we must be ready to modify or reverse them as soon as Science can advance to occupy the field in question. In other words, the work of Philosophy is evaluation and criticism.[1]

We find clear indications of the truth of the position here taken, if we observe the attitudes of some of the most enlightened of contemporary philosophers towards questions which have long been regarded as falling in the province of Philosophy, but which, being questions of fact, properly belong to Science. Thus, Professor John Dewey, confronted with the question : Does personality in any sense or manner survive the death of the body ? answers that the question is one of fact to be decided by the empirical evidence. In a similar spirit, Mr. A. D. Broad, one of the most competent and influential of the younger British philosophers, makes a comprehensive survey of the problem of the relation of mind to body, showing throughout strict regard for the empirical evidence. As a philosopher he sets out with one advantage over the mere men of science, namely, he knows that any one of the many theories may be true, except those which are self-contradictory. And, reviewing all of them, he inclines to accept that one which was formulated by T. H. Huxley under the influence of Atomic Materialism and named by him *Epiphenomenalism,* the theory that mental process is merely a shadowy by-product of the brain processes, a by-product without influence upon the

[1] Note 2, on *Differentiation of Science and Philosophy.*

course of events. And he is prevented from accepting that view only by the fact that, having, true to the empirical spirit of Science, paid, unlike most philosophers, some little attention to observations made in the field of *Psychical Research*, he finds in them some ground for believing that human personality is not wholly dissolved on the death of the body.

For the solution of the question raised by Modern Materialism, for an answer to its challenge to so much that mankind has long held to be truths of the highest importance, we can properly look neither to physical science nor to Philosophy. Here, as elsewhere, the function of Philosophy is criticism and evaluation, the clearing away of ill-based prejudices such as so often distort the mental activities of men of science. We must rather look to that branch of Science which is directly and primarily concerned with the facts of Mind, of mental life and activity, namely, psychology.

The challenge of Modern Materialism throws then, I say, a great responsibility upon psychology. To the question: Is Modern Materialism valid? Is it well-based? Is it 'an impregnable rock'? the answer must come in the main from psychology. This conclusion may be distasteful to you. For, strangely enough, psychology—the science of mental life—has acquired the reputation of being the most materialistic of the sciences, the science which, more than any other, tends to maintain Modern Materialism and to subvert all the ancient beliefs about the status and role of Mind or Spirit that are incompatible with it. This reputation, so unsavoury to most of you, is not altogether undeserved. The psychology of a bygone age relied largely upon the methods of an earlier philosophy that did not know its own limitations and presumed to undertake the work of Science in those fields which had not been cultivated by the methods of science. That older psychology solved the problem of the relation of Mind to body and erected an 'impregnable' barrier against Materialism by setting out with the assertions that each man's mental life is the activity of his soul, that his soul is a simple substance, an entity, indivisible and inextended, therefore indestructible and capable of continuing its activities indefinitely after the death of the body. That is to say, it deduced its answer to this great problem from a certain postulate which it assumed to be self-evident or axiomatic, but which in reality was a traditional prejudice that had been slowly evolved through long ages of 'philosophizing'.[1]

But modern psychology has, rightly and inevitably, become a branch of biological science. And modern psychologists have reacted against the dominance of their science by an assumption which they rightly regarded as a survival from the prescientific ages and from primitive modes of thought; they have for the most part scornfully thrown aside this assumption as a remnant of primitive superstition. And many of them have gone further:

[1] In my *Body and Mind* (1911, London and New York) I have traced the history of this way of conceiving the relation between Mind and body, showing the continuity of its development from primitive animism.

in their urgent desire to make psychology really and purely scientific, and in their natural resentment against the claims of philosophers to prescribe for them the outlines and foundations of their science, they have thrown out of the window not only the child with the bath-water, but also the bath itself; they have attempted to construct a psychology without presuppositions, without postulates, without ultimates or indefinables; and in doing so, repudiating altogether the critical guidance of Philosophy, they have put in the place of indefinables, deliberately chosen and consciously postulated, a number of prejudices which they have acquired they know not how or why. And chief among these has been the prejudice that psychology, if it is to be scientific, must work with the same presuppositions, the same postulates and indefinables, as the physical sciences. Hence, in the age of Atomic Materialism, thought was said to be but a shadow cast by the moving atoms of the brain, and personality but 'a moving show of fleeting shadow-shapes that come and go'. And now, in the age of Modern Materialism, the corresponding prejudice is that all psychological explanations must be mechanistic.

We cannot, then, hope for a decision of the issue raised by Modern Materialism from a psychology that has uncritically accepted *as its own foundation* one of the two alternative answers to this great question. We must look rather to a psychology that approaches its task cautiously and with an open mind, aware of its great responsibility, and not afraid to stand apart from the physical sciences, asserting its right to formulate its own methods and its own postulates, and leaving to the future the great task of reconciling the postulates and the formulations of the mental and of the physical sciences.

CHAPTER II

THE PSYCHOLOGY WE NEED

THE conclusion reached in our first lecture was that the psychologist must assert the independence or autonomy of his branch of science ; that he must approach his tasks with an open mind upon the great problem of the relations between mental and physical processes ; that he must claim for his science the chief part of the responsibility for working out a satisfactory statement of those relations, rather than take his stand upon Modern Materialism as an established dogma or truth, accepting the dictation of the other sciences ; and that he cannot leave the decision of this question to Philosophy. All this, the conclusion reached in our first lecture, is amply borne out by an impartial review of the history of speculation. I have made such a review in my *Body and Mind* ; it cannot be repeated in these lectures. But it seems necessary to say something more in substantiation of the claim, which to you may well seem over-bold (opposed as it is to the traditional view), that the determination of this great issue cannot properly be expected from philosophy.

Philosophy, in debating its own problems, has to accept, critically and provisionally, the findings of Science and to give due weight to them, to adapt itself to them. Though this principle has not yet been accepted by all philosophers, in practice it has prevailed increasingly as Science has grown more powerful and more confident; and in recent years many of the philosophers who have not explicitly accepted the principle have in practice conformed to it. In this connexion the utterances of those who are both philosophers and men of science are especially deserving of our attention.

When confronted with the dictum of Science that our mental life is wholly dependent upon, and without influence upon, the mechanistic processes of our bodies and brains, how have the philosophers reacted? Some have put aside this dictum with scorn as obviously absurd and incredible, and have gone about their own business, confident that some day, however distant, Science will repudiate this error and achieve some less fantastic formulation. Others have accepted it and have sought to harmonize it as best they could with their philosophy. A third group have met it with what may be called the stock argument of idealism. *Cogito, ergo sum,* said Descartes; 'I think' is the most sure, the most indisputable proposition that any man can make. If the materialist says 'the world consists wholly of matter and its motions,' he really asserts 'I think that the world, etc.' And, with various improvements and modifications, later philosophers have repeated Descartes' proposition as the most fundamental truth. Bishop Berkeley made it the foundation stone of modern idealism; David Hume the basis of his sceptical nihilism.

The argument against Materialism founded on this proposition has recently been restated by one who is both a philosopher and a psychologist and I will give it to you in his words. 'It is the business of the psychologist . . . to begin, not with views or theories . . . but with facts, indisputable facts of conscious experience . . . there can be no doubt of a fact apprehended with insight ; and facts of consciousness, whatever their nature may be, are of this kind. It can, moreover, be doubted—for it has been doubted—whether there are any such physical existents as bodies at all. But a sensation, say of red, or a pain, or a wish, cannot be doubted as an occurrence. It can be held that we have no acquaintance with real things as they exist in a real extra-thought world ; that they are only (what they certainly are for us) phenomena or appearances in our knowledge. Indeed, it must be held that, as far as physical existents at least are concerned, they have no sense or relevance for us except in so far as they are known in some way. The primordial fact, then, seems to be "knowing"—together, of course, with "feeling" and "willing" . . .' He continues : 'We cannot, then, take for granted a brain and nervous system as accounting for, or explaining, our knowing ; since brain and nervous system only exist, for us, in so far as they are known—in so far, that is, as they are mental objects and not physical ones.'[1]

This fundamental proposition of idealism is substantially correct and irrefutable, cavil as we may at any one of the thousand forms in which it has been repeated. Does it dispose, once for all, of Materialism, as some philosophers seem to suppose ?

[1] Prof. F. Aveling of the London University in the volume, *The Mind*, edited by R. G. S. McDowall (London and New York, 1927).

Can we safely draw from this premise the conclusion that mental activity is more real than physical process, or equally real with it, or in any sense independent of it, or possesses in any sense causal efficacy in the world of nature? Not at all. The materialistically inclined scientist receives this supposedly knock-out blow and comes back smiling. Not uncommonly he retorts (as does the editor of the volume from which I cite, a physiologist) by offering to compress for three minutes the carotid arteries of the philosopher, and thus, by depriving his brain of its blood-supply, to convince him of the dependence of his 'knowing' upon the chemical processes of his brain. For, during this period of deprivation, the philosopher will cease to think or to be in any way conscious. But this retort is as little a knock-out blow for the philosopher as the idealist argument was for the scientist. He may reply: 'I admit that the brain is in some sense real, and that its processes are in some sense important conditions of my normal mental functioning. In depriving my brain of its blood-supply you have gravely disturbed the conditions under which my mental activity normally goes on.' And he may resort to argument by analogy. 'See this electric glow lamp,' he may say; 'it emits a bright glow. Disrupt the material continuity of the filament, and the glow disappears. The savage would say: That proves that the glow is wholly caused by the filament, is nothing but a property of the filament. But we know that the glow is a function not only of the filament but also of a subtle energy that is in no sense a part or product of the filament, an energy so subtle it escapes the scrutiny of all our senses and remained unknown to science during long ages ; and, though since its discovery it has become of immense theoretical and practical importance, its essence remains obscure, one of the indefinables of physical science.'

The materialist in face of this reply may go further. 'I admit,' he says, 'the truth of what you say of the superior certainty of propositions about immediate experiences, such as pain and desire and thinking. But I ask you to imagine for a moment that my account of things is true. Is it not obvious that, if our powers of thinking have been evolved as the superior and peculiar functions of the matter of our highly superior brains, we could, and some of us certainly would, produce your idealist argument ? And it would be just as valid and conclusive as it actually is, i.e. it would be wholly misleading when applied as a refutation of Materialism.' To this the idealist philosophy has no reply ; and it is safe to say that it never can find from its own resources an adequate reply.

The inconclusive nature of the idealist argument is vividly illustrated by the divergent paths of those who make it their first step. 'What is matter but a shadow of the mind's own throwing ? ' exclaimed T. H. Huxley ; and, having thus proclaimed himself an idealist, he went on to construct a materialist account of the world in which the mind figured as but a shadow thrown by the brain. And even those philosophers who have gone on to elaborate an idealist philosophy on this basis have diverged widely on our central problem. Some, like the late

Bernard Bosanquet, have accepted the account given by mechanistic science together with its denial of teleological activity and its identification of mental processes with mechanistic brain-processes. Others, like his leader and colleague, F. H. Bradley, leave the question open, confessing that philosophy cannot find an answer.

The plain fact is that the idealist argument leaves you alone in your world of immediate experiences, a solipsist and universal sceptic. In order to escape from that intolerable and absurd position, you have to make or to choose assumptions or postulates of some sort. And those assumptions will largely determine the view of the world at which you will arrive. Berkeley postulated other spirits like himself, God, men and angels. Hegel postulated an Absolute, an all-inclusive Mind. Kant postulated a world of things-in-themselves behind the veil of phenomena, of appearances to our perceiving minds. Many philosophers have hurled invectives at this postulate of a world of things-in-themselves whose nature must remain for ever unknown or but imperfectly known to us: for to accept it is to admit that the task which they propose for philosophy is improperly conceived and the goal they desire for it one which it cannot attain.[1]

Science, on the other hand, accepts the world of things-in-themselves; and its whole effort is to

[1] Here we have one instance of the mistaken efforts of philosophy to undertake the work of science, to apply its scale of values to the determination of the nature of things. A world perfectly knowable would be of more value than one, by the nature of things, imperfectly knowable; therefore, say they, the world must be of the former kind.

attain such knowledge of them as may enable us to deal with them successfully, bend them to our purposes, subordinate them to our system of values. Recognizing that we cannot hope to know them in their inmost essence and detail, to be able to say exactly what they are, it seeks to understand more and more fully the course of events, how things or events influence one another and how we may influence them.[1]

The question: How may we best and most effectively influence the course of events? was the starting-point of Science; and the answer to it is the ultimate goal of Science, as well as in very many cases the immediate goal of particular scientific inquiries. And all our principles of explanation, our hypotheses and theories, are but so many guides to such action on our part.

IN PRACTICE WE RECOGNIZE NATURAL EVENTS OF TWO ORDERS

Now, when we survey the world from this point of view, it is obvious that all natural things and events fall into two great classes. There is the class

[1] In other words, the knowledge which science seeks is pragmatic knowledge; and many philosophers have yet to learn that pragmatic knowledge of the nature of things is the only kind of knowledge of them to which we can legitimately aspire. They continue to entertain impossible theories of the nature of scientific knowledge and of the criterion of scientific truth; notably the correspondence and the coherence theories of truth. The former sets up the impossible demand that our knowledge shall be in some sense an exact copy or representation of the world as it is; the latter has to regard as true every internally coherent fiction, no matter how fantastic it may be. Cp. Note 3 on *Sensationism and Neo-Realism.*

of events which we influence most effectively if we accept the mechanistic type of explanation as guide to action. And there is another great class of events, namely, the actions of men and of the higher animals, which we influence most effectively if we accept and apply the teleological principles of explanation. We may apply mechanical principles to these ; we may push and pull and apply electrical and chemical agencies ; but, if we do so under the guidance of mechanistic principles alone, we effect little. In order effectively to influence them we must recognize that their actions express motives, desires, or impulses directed towards goals ; and we must apply incentives that will evoke such motives ; we must reason, or persuade, or hold out inducements, or point out probable consequences. It is true that, in so doing, we work upon the man or animal through physical media, by sound and light and contact and so forth. But these physical media do not themselves evoke or modify the motives that operate ; it is the meaning of the words we utter, of the gestures we make, the signals we give, that evoke or modify the motives of our friend. And if you make use of material objects for this purpose, you do not use them mechanically or according to mechanistic principles ; if you hold out a piece of money or of food as a reward, you use it as an incentive of some motive for effort and not as a mechanical agency ; it is again, if you are successful, the meaning of what you do as apprehended by your friend which stirs the appropriate motive and modifies his course of action, bends it to your purpose.

In all such cases, then, the success of our efforts depends upon our thinking teleologically, upon understanding and applying teleological principles of explanation.[1]

Besides events of these two sharply distinct classes, there are events which we cannot easily and confidently regard as of either of these two classes, namely, the actions of lower organisms and the processes we observe in the organs and tissues of any living organism. These seem in many ways like the actions of a man or dog knowing what he wants and striving to attain it. Yet we can control

[1] This is true not only of the politician, the schoolmaster, the animal trainer, the sportsman, but also of the experimental psychologist, physiologist and zoologist; with the exception of a few highly sophisticated and perverted psychologists. The biologist thinks in terms of the struggle for existence, competition for food-supplies or territory, sexual rivalry, the functions of parts and processes of the organism in its total economy, how they subserve the great ends of survival of the individual and propagation of the species. The question he constantly asks and seeks to answer is: What is this for? How does this tissue, organ or process contribute to these great ends? And if he finds no answer to this queston in respect of any part or process of the organism, he calls it a 'vestigial remnant'; i.e. he assumes that at some time in the past history of the species it did have a function. It is true that we properly and profitably think of the parts of a machine in the same way; but that is because the machine embodies and subserves a human purpose. Contrast with this the way the man of science deals successfully with inorganic nature. Examining a crystal of felspar in a piece of granite, he does not ask: What is this for? What function does it subserve? He explains its presence and structure and genesis wholly in mechanistic terms, i.e. without reference to the future. So also in astronomy. We do not ask: What is the function of Venus or Mars in the planetary system? We ask merely what is the history of the system and of its parts. We may speculate on its future, but we do not need to do so in order to explain its existence and genesis.

them in some slight degree by bringing physical influences to bear; and we do not know how to influence them by appeal to motives. In fact our power over them is very small, just because we do not know on what principles to work upon them. It is true that, as we may push a man or a dog where he does not wish to go, or hinder him by wounds or mechanical obstructions, or kill him with poison-gas, so we may bring to an end, or gravely hinder and distort, by means of physical agencies the processes of this ambiguous kind.

It is the existence of processes of this third ambiguous class which has given plausibility and constant stimulus to the attempt to exhibit all natural processes as subject to explanations of one type only. Science has in the main sought to show that the mechanistic principles are valid in all cases. But there have always been those, and the tendency has grown stronger in recent years, who have held that the teleological principles are really valid for inorganic nature as well as organic, that the physical world also is fundamentally and truly mental.[1]

With this last view we are not directly concerned in these lectures. It will be enough if we can show good reason to believe that there is a realm of events within which the teleological or mental principles of explanation are valid, as valid as the mechanistic within the realm of inorganic nature; for the essence of Modern Materialism is, as we have seen, the claim to extend mechanistic explanation to all events, explaining away teleological or mental principles as at best but provisional stop-gaps where

[1] Note 4, *The Psycho-biological School.*

our knowledge is still insufficient for mechanistic explanation.

Now the events in dead bodies seem to be in the main of the same order as those of inorganic nature. It has, therefore, been argued that, since the living body seems to consist of, or to comprise, only the same materials as the dead body, its events also must be capable of mechanistic explanation. This is the doctrine of mechanistic biology and physiology.

On the other hand, it appears on the face of it that the living body is the scene of events which require for their explanation both mechanistic and teleological principles. The acceptance of such mixed principles for living organisms is the essence of doctrines commonly called vitalistic. And within the field of psychology or physiological psychology the acceptance of such mixed principles is called dualism or interactionism; for it implies the interaction of mechanistic and of teleological or mental events.

Our main task is to inquire carefully into the grounds for holding that teleological explanations are valid and necessary for the understanding and control of some part of the phenomena of nature. When we shall have found those grounds to be good and sufficient, we may turn to the question of the proper relations between mechanistic and teleological explanations, and briefly examine the claim of Modern Materialism that teleological explanations are at best merely provisional and to be replaced, as science advances, by mechanistic explanations, which alone are universally applicable. If this claim could be made good, it would follow that the mechanistic hypothesis is well founded, and that we have good ground to believe that the mechanistic account of the world truly represents the nature of all things; that, in short, Modern Materialism renders a substantially correct account of the world; that the universe is mechanistic through and through.

On the other hand, if we find good reason to believe that there are events of a kind such that the teleological or mental explanations of them cannot be replaced by the mechanistic and that there is no reasonable prospect of any such replacement, then we shall have accomplished the task proposed to ourselves in this course of lectures; we shall have found good reason to believe that mental activity is what it appears to be, what it has always been held to be by common sense, namely, an effective agency in the determination of the course of natural events, a designing and creative agency, capable in some degree of shaping the course of things in accordance with its desires, its purposes, its ideals. We shall have established the reality of teleological causation; and we may leave to the physical science of the future the problem of deciding whether mechanistic explanations will suffice for the inorganic realm.

THE NATURE OF INTELLIGENCE

We must begin the study of mental activity by explicitly repudiating an old-fashioned but still current method of describing mental life, that namely which consists in describing it as a flux of sensations or sensory elements conjoined and fused in various ways. This was the kind of psychology that naturally resulted from accepting Atomic Materialism as a perfect account of the physical world and as offering a pattern after which all science must model itself. The physical world seemed to have been successfully analysed into ultimate units, the atoms, and exhibited as clusters of atoms in various conjunctions; therefore it was sought to deal in similar fashion with the mental world, the world of pure experience. The distinguishable qualities of sensory experience were therefore called sensations and those which seemed to resist analysis were regarded as simple units or atoms. It was then alleged that all experience can be analysed into such atoms and exhibited as various conjunctions of them.

The analogy between the physical and the mental atoms was far from complete. For, whereas the physical atom was regarded as a stable, even imperishable, entity, the mental atom seemed to come into existence, endure but for a moment, and disappear, like a spark in the night. Further, the physical atom had mass and motion and therefore momentum and causal efficacy. But it was difficult to assign any corresponding properties to mental atoms, any causal efficacy or effective role in nature. Hence it was natural that such psychology, when it turned from description to explanation, regarded the mental atoms as somehow generated by special motions of the physical atoms. And this was the more plausible since sensations seemed to follow upon and to be caused by physical stimuli to the nerves and by consequent movements of atoms in the brain.

In opposition to and in revulsion from psychology of this sort, another school sets out from the idealist premise, that unshakeable premise which we have repeated in the language of Dr. Aveling. It begins with the individual's pure experience, and fearful lest it become entangled with materialistic theories and explanations, it refuses to budge beyond it, like a mule that refuses to leave its own familiar stable. Psychology of this type also will get us nowhere with our problem.

The psychology we need, both for this and many other purposes, is one that regards man as a natural object among others, one peculiarly favourable for scientific study because each of us, being himself a man, commands a source of knowledge of Man, such as is denied us in the study of all other objects. Each of us can study his own actions from without and from within. And he can study the actions of other men from without, and is justified, by the strongest of all arguments by analogy, in accepting (with due caution) the others' accounts of those actions as known immediately to themselves. The fact of this double source of knowledge, the combination in the description of human actions of observations of two utterly unlike kinds, the kind which alone is available in all other sciences and the kind we call introspection, this fact must in the end render psychology the most fundamental of the sciences, the surest source of knowledge of the nature of things; it is this fact which must make it the arbiter of the great question raised by Modern Materialism.

The argument by analogy, which justifies us in

believing the accounts of the inner side of their actions given by other men, applies, though here with somewhat less force, to the belief that the actions of the higher animals also have their inner side; and, although those animals are not able to give us any clear description of that inner side, we seem to see unmistakable indications of it when we observe them. The same argument applies all down the scale of animal life, though with diminishing force as we reach the actions of animals more and more remote from man in their constitution and modes of behaviour. And many thinkers have proposed to carry it still farther and to infer, in the light of it, an inner side of all activities, of all events, even those of the inorganic world.

This last extension of the argument raises a very speculative question, which necessarily remains and perhaps must ever remain in the region of vague probabilities or possibilities. At any rate, any answer to it is for the distant future and we shall not be further concerned with it here.

Man, then, is an organism that is able to observe systematically his own actions in their inward as well as in their outward aspects. He knows what it is to form an ideal, to desire to attain it as his goal, to strive towards it, to plan a course of action, to carry it out, modifying it in detail or in its larger features, according to the success or failure of his steps; and he knows also frustration and success.

Let us consider a single typical instance of such action. When I am in Europe I receive an invitation to give this course of lectures in this city in the heart of America. I accept it and promise to appear on a given date twelve months ahead. The reasons and motives that determine my acceptance we need not consider. The promise once given, the goal once chosen, the desire to realize it becomes a fixed resolution, a governing purpose which determines in a general way all my actions until the goal is attained and the purpose realized. I may travel eastward or westward. After weighing many considerations pro and con, I decide to travel eastward in leisurely fashion visiting many countries on the way. I choose between a multitude of possible routes and means of transport. In other words I form the outline of a plan of action directed to my goal. And, as I travel, I work out the plan and modify it in detail; always governed by the desire (which has become a fixed resolution) to attain my goal on the given date. After devious wanderings in strange places, I reach San Francisco and there buy a motor-car. I choose again between a multitude of possible routes in the light of a multitude of pros and cons; and again in the course of my journey I modify my plan many times, filling in the outline as my knowledge of the circumstances grows. And here I am. I have encountered obstructions and difficulties; but they have not arrested my course or turned me back. They might have been more serious; war, revolution, storms, shipwreck, pestilence, ill-health, thugs or the red tape of officials might have compelled me to alter my course, to diverge a thousand or five thousand miles from it, to travel westward instead of eastward, to modify my chosen line of action in innumerable ways. Yet, so long as I retained my strength of body and sanity of mind, I was bound to arrive in due time at this spot.

That is an ordinary though moderately complex instance of teleological or purposive action, action determined by the desire to attain a foreseen goal. Its execution required a vast amount of knowledge and a modicum of intelligence for the guidance of action by knowledge. It required also foresight of the goal and a fixed desire or resolution for its attainment.

It may well seem that the mere inspection of the outward aspect of such a course of action should suffice to convince any unprejudiced mind of the reality of teleological causation, to prove that here is a train of events of an order quite distinct in nature from all mechanistically explicable events. And experience of such a course of action in its inner aspect strongly bears out this view. The foresight of the goal is renewed at every point of indetermination, every point where a choice of alternative lines of action becomes necessary; and with each renewal of the foresight comes a renewal of the experience of desire for, or urge towards, the goal; this inner urge becoming especially strong and vivid whenever difficulties arise that require effort for their overcoming, effort of thinking and effort of bodily action. In all such instances of action the goal seems to exert a continued and frequently renewed and intensified attraction upon me. Yet there is no physical influence exerted on me by the goal.

Modern Materialism asks us to regard as deceptive all these outward and inward peculiarities of such action which seem to mark it off as wholly distinct in nature from mechanistic events ; to believe that, in reality, such action is mechanistically explicable in every detail and, therefore, mechanistic in its intrinsic nature. This is the claim we have to examine. If it can be made good Modern Materialism is justified. If it should appear that the claim is utterly without foundation, or that it has some slight foundation but yet is wildly improbable, if science is quite incapable of explaining or reproducing mechanistically the very simplest instances of such action, we shall be justified in concluding that such events are not explicable mechanistically, not explicable without taking into account as an essential condition, or causally efficient factor, the mental reference to the goal which seems to play so important a role.

The psychology we need must, then, regard man as an organism among others, one whose actions, inspected in both inner and outer aspects, have in all respects the appearance of instances of teleological causation. Human actions are in fact the pattern and exemplar of such causation, and it is only through our acquaintance with them that we have been led to conceive such causation. Any psychology that does not frankly recognize this is useless for the purpose of our inquiry (as it is useless also for most other purposes).

It must recognize also that all our intellectual apparatus and activities, the processes of perception, imagination, remembering, judgment, reasoning and so forth, are, all alike, steps towards action, incidents or events within a train of purposive activity that tends to issue in action. We think, perceive, remember, judge or reason for the sake of action. All such processes are links in the chain, incidents in the stream of activity which is the working out of desire into action. In our brief consideration of the nature of intelligence we must keep this functional relation to action clearly in mind.

PSEUDO-TELEOLOGY

Before coming to close quarters with our task, it is necessary to clear the ground of certain confusions which have been very detrimental to the clear statement of our problem and to straight thinking about it.

First, there is a simple verbal confusion of which many biologists are guilty. They speak of the purpose of an organ or an action. If you ask them whether they mean that the organ acts purposively or that the action is governed by desire and foresight of the end or goal, is teleological rather than mechanistic, they repudiate any such implication with righteous indignation. They mean merely to imply that the organ or event subserves a certain function in the economy of the whole; just as a lever or a cog in a machine may be essential to the working of the machine. It remains true, however, that, if we had no experience of purposive action, we should never have learnt to speak of functions, still less of purposes. Function implies organization adapted to the attainment of a goal. If we conceive the solar system mechanistically we cannot properly say that the purpose, or the function, of the sun is to attract the planets or to hold them in their orbits.

We may, and commonly do, speak of the purpose or function of a piece of a machine, of a lever, wheel or cog ; for the machine is in a sense a teleological or purposive organization. It has been designed and constructed to aid in the realization of a purpose, the fulfilment of a desire—say the purpose of turning out a printed news-sheet. Every machine embodies the purpose of its designer, and functions for the attainment of the goal desired by him who operates it. If it gets into motion by mechanical accident, as an empty motor-car may dash along the street, it is not functioning. But, so long as it is doing the work for which it was designed and contributing to the realization of a purpose, it is functioning ; as when your motor-car takes you home to dinner. It is then a mechanical link in a chain of teleological causation. Both the machine and its operation exhibit a derivative teleology, a purposiveness which derives from, supplements and extends, the purposive actions of its designer and operator. I suggest that we may properly speak of the functioning machine, and of all its operations, as pseudo-teleological or pseudo-purposive.

Now much of the debate on the validity of teleological explanation has been concerned with such pseudo-teleological processes. The question debated was : Does the world, and especially the organic realm, reveal such a nature that we may infer that it has been designed and constructed by a Supreme Mind for the realization of some purpose ?

Those who, like the Deists of the eighteenth

century, affirmed the validity of this inference attributed to the world of nature, or to the organic realm (including man) just such derived or pseudo-teleology. But their argument was gravely defective; for they assumed, without critical examination, the reality of teleological activity. If causal efficacy of purposive foresight can be demonstrated, it must be through the study of those instances with which we have the most detailed and intimate acquaintance, namely, our own purposive activities. The success of such demonstration would not show that the Deists were wrong; it might rather strengthen greatly their argument: for it may well be that Man is both an instrument and an agent; both a link in a chain of designed organization, and a real teleological agent, capable of furthering or of hindering the working out of the Great Design.

Similar considerations apply to the arguments and conclusions of many vitalists, the biologists who are convinced that the events of the world of life cannot be mechanistically explained. They have proposed to introduce a variety of terms implying the operation in living organisms of teleological agencies that modify or guide the mechanistic processes of their tissues. Such are the 'vital force' of the older vitalism, the '*élan vital*' of Bergson, the 'entelechies' and 'psychoids' of Driesch, the 'ortho-genesis' of Eimer, the 'organizatory factors' of Seba Eldridge.[1]

[1] The latest proposal of this sort would seem to be that of Prof. E. Rignano, made before the psychological section of the British Association at Oxford in 1926; he proposed to postulate in living organisms a peculiar form of physical energy that operates teleologically.

To any and every such proposal we may reply: There is only one form of teleological causation with which we are directly acquainted, namely our own purposive activity. Unless we can show that it has causal efficacy in the world of nature, we are not justified in assuming the reality of any such agency or attributing other than provisional validity to any teleological explanation whatsoever. And, if we can establish its causal efficacy, we shall probably find that all instances of teleological causation are of the same kind, are manifestations of mental activity of like nature with our own.

To repeat in other words: Our own actions seem to express foresight, desire and purpose, or to be teleological; if they can be explained mechanistically we have no ground for believing them to be instances of teleological causation and very little ground for believing that teleological causation occurs or has occurred at any time or place.

CHAPTER III

ACTION AS INTELLIGENCE AND AS PURPOSE

IT is only by abstraction, permissible but dangerous, that we separate intelligence and purpose, and consider them apart. Either one is inevitably falsified if described apart from the other, the complementary, aspect of all mental activity ; for all intelligent action is purposive, and all purposive action is more or less intelligent. This truth is recognized in most of the many current definitions of intelligence. Nearly all such definitions amount to saying that intelligence is the power of adapting action to novel circumstances, or is that which is manifested in such adaptation of action. In so far these many current definitions are good ; but most of them carry us no farther, they leave the essential nature of intelligence obscure.

The academic psychology of some decades ago was content to treat the intellectual life as a mere mental representing of objects or events, the having of ' ideas ' of them in consciousness ; and it was inclined to overlook even the relation of the so-called ' ideas ' to their objects and to describe the life of the mind as a mere stream of ideas, a phantasmagoria without function, a dance of shadows subserving no purpose.

Fortunately, psychology is beginning to make progress ; and from various directions come useful suggestions as to the intimate nature of intelligence.

First and most important is the recognition that intelligent activity implies not merely representations of objects or events (as pictures on the screen are representations) but also a mental reference to the object represented ; it involves awareness of the object, thinking of the object, or, as we may conveniently say, thinking the object.[1]

Here the mechanistic account of mind encounters at once a great difficulty. So long as mental process was regarded as nothing more than a succession of pictures of objects, each picture a mosaic of sensations, it was plausible to regard such pictures, such representations or reflections of objects, as analogous to other reflections and representations. The still pool reflects the tree that stands beside it. The mirror reflects the stream of objects that pass before it. But we have not the slightest ground for believing that the mirror is in any sense or degree aware of the man whose face it reflects. Between the man and the mirror a cognitive relation obtains ; but it is a one-sided relation. Each reflects the other ; and, in so far, a mechanistic explanation of both reflections is plausible. But the man, in addition to reflecting the mirror, is aware of it. He is active in that unique fashion which we call knowing or thinking the object ; consequently

[1] The object may be of any conceivable kind, physical or mental, imaginary or real, abstract or concrete, general or particular. For the sake of simplicity we may keep in view action in relation to physical objects. It is convenient, though a little unusual, to extend the usual meaning of the term ' thinking ', so as to make it cover every form of awareness of objects. The word ' knowing ' implies knowledge or true thinking ; but much of our thinking is illusory or fallacious and hence is not knowing.

he changes or develops in and through the activity, as shown by the fact that afterward he can remember the mirror. The mirror on the other hand remains passive. Neither at the moment of reflection nor later can the most refined investigation discover in the mirror any evidence of activity or change involved in the reflection.[1]

To think an object is, then, to enter into this

[1] It may seem absurd to insist on this difference between the two reflections, the mirror's and the man's. But it is necessary in view of modern efforts to show that thinking is not a unique mode of activity, but merely a special case of reciprocal influence such as is involved in physical processes, or that the cognitive relation is only one among many kinds of relations of the same order. Thus some philosophers tell us nowadays that the cognitive relation is a special case of the causal relation. And Prof. S. Alexander, in the interests of an extreme form of neo-Realism, asks us to believe that when the mirror lies upon the table, the mirror knows the table and the table knows the mirror. This view is of importance in relation to the theory of emergent evolution, and we shall have to return to consider it further in that connexion. Here I will only insist that the unique nature of the cognitive relation appears clearly if we reflect that all other relations of the physical object are constitutional, or constitutive of the object as it exists at any moment. We cannot completely describe the object without describing all its relations other than the cognitive. But the physical object remains indifferent to our thinking of it. At one moment a million men may think of an aeroplane whirring its way across the Atlantic Ocean, and a hundred thousand men may perceive it as it departs or arrives ; but all this thinking of it, this multitude of cognitive relations, makes no difference to the aeroplane, except in so far as the thinking prompts and directs human action upon it. All other relations are reciprocal ; the cognitive relation is not, it is purely one-sided. This fact marks it as unique, as of a different order from all others.

Idealists also ignore or deny this uniquely one-sided nature of the cognitive relation when they say that the mind constitutes its objects in thinking them. This is one form of the ever-renewed attempt to confuse the object of which we think without thinking of it. Yet to keep this distinction sharp and clear is the prime condition of all effective scientific thinking.

unique relation to it, and to be active about it, without acting on it. And we have no faintest indication that any objects other than living organisms ever can or do acquire this unique relation to other things or exercise any activity remotely similar to the activity of thinking.

Of late years much emphasis has been laid upon the symbolic functioning of the mind. It is recognized that the symbolic function pervades our mental activities of all levels. All use of language is symbolic activity; and language is involved in all higher thinking. But the same is true at the level of simple sense-perception. When you see a red rose, its coloured shape is all that you immediately apprehend; but you accept it as signifying, or as symbolical of, the rose. The association psychology of the past attempted to explain the fact by saying that the sensations of colour have become associated with those of odour, touch and so forth, previously experienced on perceiving roses; that these therefore are reproduced when you see the rose; and that this cluster of sensations, those directly excited and those reproduced, constitutes your thinking of the rose (or your percept, your mental representation of it).

There is a certain amount of truth in this account; but it is very imperfect. The perceiving of the rose is far more than the coming into existence of a cluster of sensations. You become aware of it as a physical object and a living object; and you treat it accordingly, implicitly anticipating all the properties which you have learnt to belong to such objects.[1]

[1] For a fuller statement of this complexity of the perceptive process see my *Outline of Psychology*.

The addition of reproduced sensations to those immediately excited does not convert the cluster of sensations into an awareness of the object. The awareness of the object, the objective reference is primordial in mental activity.

The gist of the matter has been excellently stated in a recent lecture by Prof. A. N. Whitehead.[1] 'Coloured shapes seem to be symbols for some other elements in our experience, and when we see the coloured shapes we adjust our actions toward those other elements.' But, he continues, 'The familiar language which I have used in speaking of the "projection of our sensations" is very misleading. There are no bare sensations which are first experienced and then "projected" into our feet as their feelings, or on to the opposite wall as its colour. The projection is an integral part of the situation, quite as original as the sense-data. It would be just as accurate, and equally misleading, to speak of a projection on the wall which is then characterized as such-and-such a colour.'

He goes on to say that, of the other elements of experience symbolized by the immediate sense-data, the most important is experience of the causal efficacy of the thing perceived. Now this experience of the causal efficacy of things is not merely experiencing some particular sense-quality; it involves an awareness of the object, that reference to objects which is the primordial and unique function of the thinking process.

Let us consider very briefly the testimony of that new school of psychology which is now ascendant in

[1] *Symbolism, its Meaning and Effect* (N.Y., 1927).

the German universities, the school of *Gestalt* or configuration. The Germans are very thorough, and they move slowly ; but they do move. And this school has moved so far from the academic psychology of its predecessors that it insists upon the fact of ' insight ' in all intelligent action. And it enforces this by the most careful and detailed observations of the behaviour of men and animals. When man or animal behaves intelligently, he gives evidence of something more than the experience of some cluster or pattern of sensations ; over and above all such mere evocations of sense-qualities is the attainment of ' insight '—it may be at the moment the sense-impression is received, it may be later.

What, then, is this insight ? The members of this school, who so rightly insist upon and vividly demonstrate the reality and importance of ' insight ' as something quite other than and additional to the experience of sense-qualities, nowhere undertake to tell us in what it consists. There can, I think, be no doubt of the answer to this question. ' Insight ' is the grasping of, or intuition of, relations more especially relations of time, space and causality.

The same conclusion has been reached by Prof. C. Spearman, who, with the aid of a number of co-workers, bases it upon a vast amount of careful experimental investigation of the nature of intelligence and upon the mathematical handling of the data of observation.[1] Intelligent action, he tells

[1] His observational basis cannot be presented here even in the most sketchy way. I refer you to his two important volumes, *The Nature of Intelligence and The Principles of Cognition*, and *Abilities of Men* (London, 1923 and 1926).

us, always implies an activity which he calls the 'eduction of relations'.

Still another psychologist, Mr. Claremont, has recently put forward independently the view that the essence of intelligence is the grasping of causal relations.[1] While yet another has put up a strong argument in favour of the view that the essential function whose range or degree is measured in imperfect fashion by all the many contemporary efforts to measure degrees of intelligence is 'confluence', the bringing together of many data in the one highly complex synthetic activity that guides intelligent action.[2] This synthetic function, this confluence, is certainly very real and important; and it involves the synthesis, not merely of many sense-qualities, but also of many relations.

Psychologists, after neglecting, one may fairly say ignoring, relations, are thus beginning to take account of them, to see that all intelligent action implies appreciation of relations. It is true that William James did not entirely overlook them; and that, earlier still, Schopenhauer insisted that all intelligent action, whether of men or animals, implies appreciation of temporal, spatial and causal relations. But the difficulty of reconciling this fact with the sensational type of psychology that comports most readily with a mechanistic scheme of explanation has obscured it for the great majority of modern psychologists.

Even at the present day many of them will be shocked, almost horrified, by the statement that

[1] *The Nature of Intelligence* (London, 1927).
[2] Victoria Hazlitt, *Ability* (London, 1926).

animals in perceiving things and acting intelligently in relation to them show good evidence of appreciation of relations of time, space and causal efficacy. Yet a little unprejudiced observation and reflection should suffice to convince anyone of its truth. It cannot be denied that the higher animals display intelligent anticipation; as when a dog whines outside the door at which he wishes to be admitted; or grows joyfully excited and rushes to the door when he hears in the distance his master's voice or footstep. But such intelligent anticipation involves appreciation of temporal sequence and relations.

That animals, and even animals comparatively low in the scale of organization, appreciate spatial relations is very clearly shown by the wide distribution among them of the power of returning home, of finding their way back to the breeding place, the nest, hive, burrow or lair, or the place where food or water has been found. Many of the more astonishing of these performances remain obscure; and this fact lends a certain air of plausibility to speculative suggestions of mechanistic explanation of them.[1] But it is clear that in many such instances the animal guides his course by the aid of recognized landmarks. Now mere recognition may be plausibly brought within the mechanistic

[1] As when it is suggested that an unknown form of physical energy radiates from the spot to which the animal returns, directly influencing its movements. In my *Outline of Psychology* I have examined such attempts and shown that they are utterly inconsistent with the facts and that in many cases the animal's return home implies appreciation of spatial relations similar to our own. In a recently published book, *How Animals Find their Way About*, Prof. Rabaud, of Paris, surveys all the evidence and reaches the same conclusion.

scheme by saying that it is merely the addition or complication of the 'percept' by a feeling of familiarity, and that the behaviour that implies recognition is merely the expression of a motor habit. But if the animal is to guide his movements by aid of the landmark, he must do more than recognize it and more than react to it in some habitual fashion. And it is clear that he does do more. If he is merely wandering in search of food, or is going out from home on some quest, he does not react to the landmark as he does when returning home. What in the light of analogy we may fairly call his desire or his purpose to reach his home, or, in non-committal terms, his set towards his goal, determines his mode of reaction or recognition of the landmark. And he can guide his course by aid of the recognized landmark only if he has appreciated and learnt its spatial relation to his home.

Animals in general seem to have much more developed appreciation of spatial than of temporal relations; but the evidence in both cases is indisputable. The evidence of their appreciation of causal relations is perhaps less obvious, and will be more obstinately resisted by psychologists dominated by the mechanistic prejudice. Yet such evidence is afforded when animals learn to deal effectively with mechanical obstructions, such as latches, bolts and bars. The school of animal psychology that has been dominant in America seeks to explain all such actions as the mere repetition of movements made in the first place in a *purely random* fashion, i.e., as incidents in a series of aimless and undirected non-intelligent movements. In many instances this view may appear plausible ; yet careful observation and consideration of almost any instance will show that it is untenable. The best instance known to me as illustrating the intelligent nature of such action and the appreciation of causal relations,[1] is one described by my son.[2] He taught white rats to obtain their food by opening a box, the lid of which was fastened by interlocking latches. As the rat mastered the problem presented by each latch, a new one in a new position and new causal relation to its predecessor in the series was added. In this way the rats learnt to open a series of fourteen interlocking latches. Now the most significant fact in the present connexion was that, after learning to deal effectively with some half-dozen latches through a long course of instruction, the rats very quickly mastered the remaining members of the series without help or instruction of any kind. They showed, that is to say, an appreciation of the causal relations of the latches and of their own actions upon them, an appreciation which rapidly improved through repeated experiences of dealing with such problems.

This view, that animals do appreciate causal relations, may be supported by further citations from the lectures of Dr. A. N. Whitehead :[3] ' The world, given in sense-presentation, is not the aboriginal experience of the lower organisms, later

[1] Apart perhaps from certain instances provided by Prof. Köhler's descriptions of the behaviour of chimpanzees.

[2] In an article in *Journal of Comparative Psychology* (1927), ' Notes on Instinct and Intelligence in Rats and Cats.'

[3] Op. cit.

to be sophisticated by the inference to causal efficacy. The contrary is the case. First the causal side of experience is dominating, then the sense-presentation gains in subtlety' (p. 49). 'Pure instinct is the response of an organism to pure causal efficacy . . . pure instinct is the most primitive type of response which is yielded by organisms to the stimulus of their environment.' Again, 'the mode of causal efficacy is the experience dominating the primitive living organisms . . . it is a heavy primitive experience.'

And Dr. Whitehead shows very clearly also that appreciation of causal efficacy is primitive in our own experience, rather than an intellectual inference which we make only after much acquaintance with things and their ways. 'The perception of conformation to realities in the environment is the primitive element in our external experience. . . . Our primitive perception is that of "conformation" vaguely, and of the yet vaguer relata "oneself" and "another" in the undiscriminated background.' Experience of this type 'however insistent, is vague, haunting, unmanageable—is heavy with the contact of the things gone by, which lay their grip on our immediate selves. . . . Those periods in our lives—when the perception of the pressure from a world of things with characters in their own right, characters mysteriously moulding our own natures, becomes strongest—those periods are the product of a reversion to some primitive state. Such a reversion occurs when either some primitive functioning of the human organism is unusually heightened, or some considerable part of our habitual sense-perception is unusually enfeebled. Anger, hatred, fear, terror, attraction, love, hunger, eagerness, massive enjoyment, are feelings and emotions closely entwined with the primitive functioning of "retreat from" and of "expansion towards". . . . But "retreat from" and "expansion towards", divested of any detailed spatial discrimination, are merely reactions to the way externality is impressing on us its own character. You cannot retreat from mere subjectivity; for subjectivity is what we carry with us. . . . These primitive emotions are accompanied by the clearest recognition of other actual things acting upon ourselves. . . . When we hate, it is a man that we hate and not a collection of sense-data—a causal efficacious man. . . . There can be no useful aspect of anything unless we admit the principle of conformation, whereby what is already made becomes a determinant of what is in the making. . . . In practice we never doubt the fact of the conformation of the present to the immediate past. It belongs to the ultimate texture of experience, with the same evidence as does presentational immediacy. The present fact is luminously the outcome from its predecessors, one quarter of a second ago.' And 'Causal efficacy is the hand of the settled past in the formation of the present. . . . So far as concerns the causal efficacy of the world external to the human body, there is the most insistent perception of a circumambient efficacious world of beings.' [1]

[1] I particularly welcome this impressive statement by one who carries great weight both as philosopher and as man of science, because it bears out so fully the unorthodox account of perception and of the conditions of our belief in the reality of things given in my *Outline of Psychology*.

The animal, of course, does not conceive temporal, spatial and causal relations abstractly ; he cannot talk about them. He merely experiences them and adapts his actions accordingly ; and that is the essence of intelligent action. And in a similar way the young child experiences, appreciates, such relations long before he becomes able to think of them abstractly, to judge them explicitly or to reason about them.

Now, so long as intelligence or mental life is described as a mere stream of sensations or of ideas or of representations, there is a certain plausibility in the view that it may be explained mechanistically. It is true that we do not know, and probably never shall know or be able to explain, how a physical stimulus to a sense-organ or nerve evokes or results in a sensation, an idea, or a representation. But given an empirical correlation between sensations (or other details of consciousness) and brain-processes of certain types, it might be possible to explain mechanistically the succession of the mental facts by explaining on that principle the succession of the corresponding brain-processes. And that is the sort of mechanistic explanation of mental life which, it is claimed, is in principle possible ; and that claim is of the essence of Modern Materialism.

But, when we have recognized that the essence of our mental life, in its aspect as intelligence, is not a mere succession of sensations, ideas, or representations of things, but a thinking of things and of their spatial, temporal and causal relations, and that such awareness of relations plays an essential part in the guidance of our actions, the mechanistic assumption at once appears far more extravagant.

The causal relation is not an obvious and striking feature of natural events, not one that, like the physical properties of things, might be expected somehow to impress itself on the organism ; it is not itself a causal agency. It is so obscure a feature of physical reality that, as we have seen, some philosophers deny altogether its reality or objectivity, asserting that it is merely a way in which we falsely conceive things to be related. Yet David Hume's attempt to display our thinking of causation as the consequence merely of the repeated experience of certain successions of impressions (and to be in fact nothing more than a revival of such successions in the mind) is admitted on all hands to have failed completely.

We believe we discover causal relations in the physical realm. We believe that our thinking of those relations has causal efficacy in guiding our actions as we manipulate physical things. And both beliefs are verified in the completest possible way, namely, pragmatically by the continually renewed successes of both beliefs in leading us to our desired goals. And be it noted that the verification of the second belief is at least as complete as that of the former. The two beliefs must stand or fall together ; for they have the same evidence and verification. Indeed, the second belief is the more fundamental ; the former derives from and depends upon it. Our belief in causal relations in the physical realm is verified only by the success of our thinking of them as guidance to our own causal intervention in physical events. Yet Modern Materialism accepts the former belief as true and infers from it the falsity of the second. It is, thus, by its very nature self-destructive and therefore logically absurd.[1]

THE PURPOSIVE ASPECT OF INTELLIGENT ACTION

We have found that to deny causal efficacy to our thinking considered in its aspect of intelligence is logically absurd. We have now to consider human action in its purposive aspect; and, since we have already established the causal efficacy of the thinking that guides such action, we may now concentrate our attention on the further question: Is the causal efficacy of such activity of the same general order as that commonly attributed to physical events, namely, is it mechanistic, or is it of a radically different order?

Is there any kind of observation or of reasoning, or of observation and reasoning, that can yield a sure answer to this question? How can we attack the problem with good hope of finding an answer?

First, we may examine typical instances of purposive action, asking whether there are any peculiarities that are common to them but not found in instances that seem to be mechanistic.

Secondly, having clearly defined these peculiarities, we may critically examine the more noteworthy of the attempts that have been made to describe and explain purposive actions mechanistically, asking whether any of these are successful, whether the

[1] Note 5 on *The Role of Meaning in Action*.

peculiarities which seem to mark off purposive action as radically different from mechanistic can be accounted for in mechanistic terms ; whether, in other words, they can be explained away.

Thirdly, we may examine the grounds which lead so many men of science to assume that purposive or teleological action must be and in principle can be explained mechanistically.

PECULIARITIES OF PURPOSIVE EVENTS

Those of us psychologists who recognize that all action expressive of Mind is purposive, in the broadest sense of the word, are represented by the mechanists as postulating entities of a peculiar kind which we call 'purposes' and to which we ascribe what our critics are pleased to call a 'mystic potency'[1]. This, of course, is to misread us and to approach the problem of purposive action in a way that is prejudicial to all understanding.[2] We need not pause to try to define exactly how the substantive 'purpose' may properly be used. Let us

[1] They assume that our purposive psychology differs from their mechanistic psychology only in adding to the factors which they postulate (namely, the mechanistic interplay of ideas or the neural correlates of ideas) this other factor of 'mystic potency', a purpose ; and in their estimation our view is manifestly false because no one is able to give even the vaguest description of this supposed entity, 'a purpose'.

[2] In a similar way the sceptics, the solipsists, the pyrrhonists, all those who profess to believe that science has no business to try to explain and should be content to describe, and perhaps to state certain coefficients of correlation, incline to accuse those who believe in physical causation of postulating mysterious entities called 'causes' or 'forces' and of ascribing 'mystic potency' to them.

rather begin by repeating the statement that the purposive and the intelligent aspects of human action can be separated in thought only by abstraction, that they are always complementary coexistent aspects, the one more prominent in some instances, the other in others.

We have already dwelt upon a striking instance of a purposive train of action, action that is clearly a striving towards a goal, namely, action consisting in locomotion towards a particular place which we desire to reach. That such action is intelligent, that it implies guidance through appreciation of the nature of things, and especially of their temporal, spatial and causal relations, is clear. The principles involved are not essentially different in those instances in which our goal is separated from us, not by space, but by time only,[1] or requires for its attainment some physical construction or some purely mental activity or construction, as in solving an arithmetical problem.

Bodily locomotion toward a goal may serve as the type of action most clearly exemplifying purpose in both animals and men. Some animals, in their migrations, display such movement over distances comparable to that of our illustrative journey half round the globe. But, since our knowledge of such migrations remains very imperfect, we may dwell upon simpler, more familiar instances.

Consider the case of a man or dog, at an upper window in a large house, who sees or hears a friend outside beckoning or calling him. He shows signs

[1] As when we rummage in memory, seeking to recall forgotten details.

of excitement and runs downstairs and to an outer door. He finds the door fastened; he turns and runs to another, perhaps at the back of the house, runs through it and round the house and so to his friend.[1] The difference between the behaviour of the man and of the dog is chiefly that the dog is apt to show more of *relatively* random and ineffective movement, more of vaguely or ineffectively directed trial movements. The principles are the same if the action is not initiated by any perception of the goal; as when, for example, the man or dog is released from captivity ten miles from home and forthwith proceeds to his home.

Such actions are purposive and they are teleological. In some sense the goal plays a part in determining the action. Now in discussing teleological causation it is usual to introduce an unnecessary air of mystery by saying that in such cases the end causes or determines or initiates or governs the action.[2] The only teleological causation with which we have any intimate acquaintance is our own purposive action. In all such cases we see clearly that it is, not the goal as such, but the thinking of the goal that plays this causative governing role. The goal may be one we falsely conceive; or it may have been destroyed or removed from the place where we conceive it; but that does not make any

[1] Cp. the description of such behaviour of a dog given by Prof. L. T. Hobhouse in *Mind in Evolution*.

[2] The statement of the case is made only a little less mystifying if we say that 'the idea' of the goal causes or determines or governs the train of action; for 'idea' implies an entity of which no clear, consistent and useful account has ever been given or can ever be given.

essential difference. The man may go home desiring and confidently expecting his dinner; but the kitchen range may have gone wrong. The dog may go home, seeking his accustomed bone; but the cupboard may be bare.

The end or goal of a train of action involves some relation not yet obtaining; since then the goal does not exist in the present, it cannot be a causal agency in the events that lead to its realization; and 'an idea of the goal' (since we cannot conceive 'an idea' in any clear fashion) is not easily conceived as a causal agent. But thinking is an activity with which we all have intimate and immediate acquaintance; and that such activity should have causal efficacy seems so natural and inevitable that it can be doubted or denied only by those who have undergone a long course of perverting sophistication. But it is said: How can mere thinking, even thinking of a goal, make a difference to the course of physical events? We answer: It is not mere thinking that makes the difference; it is rather thinking in that peculiar way we call desiring, or being averse from, that of which we think. 'There is nothing either good or bad, but thinking makes it so.' That of which we think we may think to be good or bad; we may think of it as desirable, to be ensued, to be striven for or towards; or we may think of it as repugnant, as odious, as something to be destroyed, avoided or obviated. And it is thinking of this sort that makes a difference; it is thinking of this sort for which we claim causal efficacy upon the course of events.

Such causal efficacy cannot be denied. It is

manifested in and illustrated by all the multitude of physical constructions, art-productions and social institutions that make up what we call our civilization or culture, what the Germans call *Geist* or objectified spirit. The total of such culture is the product of a multitude of creative efforts, and of positive preferences and choices of those things that have seemed good ; and of an equally great multitude of rejections and repudiations of things that have seemed bad. In some cases the existing state of affairs has resulted from an immense number of vague slight preferences, desires or aversions, coinciding in direction, co-operative and summative in effect. In other instances some one man has conceived some state of affairs that does not exist and never has existed ; and it has seemed to him so supremely good or desirable that he has said : 'Henceforth I concentrate all my energies and devote my life to making this ideal a reality.' He has held fast to his purpose and in the end has prevailed, has made real or actual that which he had conceived as possible and desirable.

The causal efficacy of such thinking cannot then be denied, is not seriously denied. But it is the essence of Modern Materialism to maintain that such thinking is causally efficacious only in the same way as the events of the inanimate world, that is to say, mechanistically, or without taking into account that appreciation of relations which is the essence of intelligence and that valuation, that striving towards or away from, which is the essence of purpose.

The great question, then, is : Is the causal efficacy of value-thinking, of striving, of intelligently directed goal-seeking, of the same order as that causal efficacy which alone is recognized by the physical sciences and invoked by them for the explanation of all physical events ? [1] We shall find that we cannot reduce such teleological causation to a special case of mechanistic causation, cannot even intelligibly describe any particular instance of purposive action without taking into account the awareness of the goal.

Keeping close now to our type-instance of purposive action, the travelling towards a distant goal, we notice that the goal seems to act like a magnet upon the agent, or to exert upon him some attractive force comparable to gravitational attraction. These analogies are so striking that they have passed into poetry and common speech.[2] We say that a man's home attracts him, or is his magnet or loadstone ; or we say that he gravitates towards his home (or other centre of attraction). And the mechanists, when they attempt to explain the more striking instances of such behaviour in animals, commonly resort to the postulation of some physical influence radiating from the home, something comparable to magnetism or gravitation, and postulate also various purely hypothetical organs in the animals for the reception of such influences. In our ignorance of animal migrations we cannot positively assert that

[1] At this stage it seems well to point out that, if we should find overwhelming reasons for denying the possibility of mechanistic explanation of such causation, it will still remain possible that all causation may be shown to be of one order only ; but this would involve the radical transformation of the current ways of conceiving physical causation and the assimilation of all physical to mental or teleological causation.

[2] 'Even the weariest river winds somewhere safe to sea.'

in no case is there any such physical guiding influence. But we can assert with the utmost confidence that in all ordinary instances of returning home (including the vast majority of such instances of animal behaviour) there is no such physical guiding influence at work.

The implied analogy between ' homing ' on the one hand and magnetic or gravitational attraction on the other is a false and misleading analogy. A motor boat steered by a magnetic compass presents no true analogy to the behaviour of the homing man or animal. In the former case a physical influence from outside steers ; in the latter, the steering apparatus is wholly within the organism and working independently of all physical influences from outside.

Let us say that the ' homer ' has imprinted within him some picture of the home. Clearly that is not enough : it must be a picture of the home set in the midst of a map of the locality, including the whole of the locality through which the journey is made, a map representing all the more essential spatial relations of the home to the locality. We can imagine, perhaps could construct, a motor-car which, being provided with such a map, would steer itself along a network of roads and, always taking the appropriate turnings, would arrive at the designated goal. But suppose that, after this ingenious mechanism (automatically steered by a map within it) has started on its way, one of the roads it must take is destroyed or barred by insuperable barriers : the mechanism fails utterly. And equally, if you divert it however little from its prescribed course, force it to take one wrong turning, the whole process fails. The homer on the other hand, though he may be temporarily baffled by such difficulties, eventually overcomes them. You may bar any or all of his possible roads at any point, and he will make detours or in some way break down or surmount your barriers, so long as that is physically possible. His detours may lead him far afield, in circles, or directly away from his goal. Or you may turn him aside with false information or threats or the offer of a meal; yet in every case he rectifies his course eventually and, in spite of all obstacles and false turnings, arrives.

Note this fact also: the more serious the obstacles, the more numerous the false turnings and misdirections, the more does his output of energy increase; until perhaps it becomes, as we say, frantic; thus to continue until either the goal is reached or physiological exhaustion supervenes.

Three features of such an instance of homing are highly peculiar, over and above the absence of all physical influence from the goal. First, the infinite adaptability of the course of action. I say this adaptability may properly be called infinite: for there is literally no limit to the number and variety of obstructions and difficulties that might be interposed between the homer and his goal and which he would successfully circumvent or overcome, displaying in doing so appreciation of temporal, spatial and causal relations. We may imagine a machine that would adapt itself to a number, a strictly limited number, of such obstructions; but not one that would circumvent or overcome any one of an unlimited number and variety.

The second highly peculiar feature of such action is its persistence (under which comprehensive term we may include the spontaneous renewal of action after arrest and augmentation of action in face of obstruction). You may incarcerate your homer for a day or a year or ten years; and, as soon as the physical restraint is removed, he resumes his homeward trek. Or you may reduce him to inanition by a blow on the head or a sufficient dose of chloroform; and again, as soon as the physical arresting influence passes away, he resumes his journey. Or you throw yourself across his path and say he shall not pass; and he struggles furiously against you. In the inanimate world there seems to be nothing strictly analogous to this increase of output of energy in proportion to and in excess of the resistance offered. The more you compress a steel spring, the more strongly it resists your push; but in this case you are transferring energy to the spring as you compress it. It returns at most a thrust just as strong as that which you apply to it. The animal or man striving towards a goal appreciates the magnitude of the resistance opposing his effort and puts out sufficient energy to overcome it (within, of course, the limits of his capacity). We see this peculiarity of purposive action illustrated most simply when one puts out one's hand to lift an object of unknown weight; but it is common to all forms of purposive action.

The third peculiarity of purposive action, as a manifestation of energy or of energy transformation, is that it ceases as soon as its goal is attained. Here again we seem to have something without parallel in the inanimate world.

These three peculiarities of the energy manifestations of purposive action considered purely objectively appear in very different degrees in different instances. We regard them as objective indications of the strength of the impulse or desire that moves the organism to action, of the intensity of the effort made by it.

We gain further understanding of these goal-seeking actions when we examine and reflect upon our experiences in the course of such actions. We know what it is to be moved by a feeble or a strong impulse, to desire faintly or strongly and intensely, to make a slight or a great effort. These are experiences of conation or of willing in the broadest sense of that word. We are here confronting the question: Can the phenomena of conation or Will be mechanistically explained?

If it can be shown that the simpler instances of conation can be so explained, it will follow that the highest and most complex instances of human volition are susceptible of similar explanation. We cannot hope to save the freedom and autonomy of human volition by arbitrarily setting it apart as something altogether distinct from the lower forms of action, something of a radically different order. We may hope to achieve this only if we can show that the lower kinds of action also exhibit, in however humble form, something of what we claim for human volition: for the latter is a complication of and a development from the former, a complication and development which each one of us achieves in proportion as he develops character and intellect.

When we examine purposive action in the light of

our own experience of it, we note first that it involves the unique, the altogether peculiar, experience of activity, whether it be in the form of an obscure impulse, a well-defined desire, a self-conscious and deliberate decision, or a bodily or mental effort of the most explicit kind. In all such experiences there is this core of sameness, the sense of activity, of doing, or of striving to do, something. And in and by our successful efforts or strivings we achieve the belief in our own causal efficacy.[1]

There is the strongest reason to believe that our belief in the causal efficacy of the things about us (which, as Whitehead so well shows, is fundamental and primitive in sense-perception) is correlative with and dependent upon such experiences of our own activity. If we imagine a man devoid of all such experience, one who passively received all the physical influences that rained upon him, we can see that he would be a natural solipsist; he would never perceive or believe in the causal efficacy of either himself or of other things and persons. The world about him would appear as a mere shadow-show, a phantasmagoria of sense-phenomena; and, in so far as he might attempt any scientific account of it, it would be a purely descriptive science with, at best, statistical correlations in place of causal explanations.[2]

[1] A persistent tradition in modern psychology will have it that this experience is nothing but a peculiar kind of sensory experience, namely, that excited by the working of our muscles. The grounds for rejecting this view are conclusive. They are set forth in my *Outline of Psychology*.

[2] In my *Outline of Psychology* I have shown that our belief in the reality of ourselves, of other persons and things, derives from

Such experiences play a fundamental and essential role in leading us to conceive of forces, energies, activities of any and every kind. When, then, Modern Materialism seeks to explain all purposive action mechanistically, it is repudiating its own foundation, the very ground of our belief in causation or causal efficacy of any kind ; it presents itself as in the act of sawing off the branch on which it sits ; and, if it should be successful, it must fall to the gound in ruins together with all its system of causal explanations.

We sometimes speak of blind impulse ; but the expression must not be taken literally. It is doubtful whether impulse is, in any instance, altogether blind. It is more probable that in its lower instances it is only relatively blind. I repeat that purpose and intelligence are two aspects of action : we cannot find instances of purely impulsive action ;

and is based upon such experiences of activity and of the resistances offered to our efforts to change things.

Occasionally we lapse into this solipsistic attitude towards the world. Such an occasion is well illustrated by Whitehead in the following passage : ' Sometimes men are overstrained by their undivided attention to the causal elements in the nature of things. Then in some tired moment there comes a sudden relaxation, and the mere presentational side of the world overwhelms them with the sense of its emptiness. As William Pitt, the Prime Minister of England through the darkest period of the French revolutionary wars, lay on his death-bed at England's worst moment in that struggle, he was heard to murmur, " What shades we are, what shadows we pursue ! " His mind had suddenly lost the sense of causal efficacy, and was illuminated by the remembrance of the intensity of emotion which had enveloped his life, in its comparison with the barren emptiness of the world passing in sense-presentation ' (op. cit.). Something similar happens in those psychasthenic patients whose vital energies are sapped by internal conflicts, as I have pointed out in my *Outline of Abnormal Psychology*.

even the most stereotyped forms of instinctive behaviour, such as we find in the insect world, seem to be intelligent in a lowly fashion. We cannot, then, expect to interpret or understand any purposive action without taking into account its intelligent aspect. We have to take it into account in trying to understand the three great peculiarities of purposive action stated above.

The variability or adaptability, the re-enforcement by opposition or obstruction, the persistence until the goal is attained, and the cessation upon attainment, all these objective peculiarities mark action as goal-seeking and imply foresight of the goal as well as the urge towards it. Mere representation of the goal is not enough ; there must be, and in all the purposive actions we can inspect from within there is, forward temporal reference ; the goal is thought of as lying before us in time. Beside this appreciation of its temporal relation, there is, in all action involving bodily activity, appreciation of spatial relations ; and in action involving manipulative activity there is appreciation of causal relations.

Action observed objectively implies these intelligent aspects ; and in those actions which we perform with sufficient deliberation to permit of full inner inspection we are able to report these forms of awareness.

Inspections of teleological action from within and from without alike imply intelligent guidance of impulse. We have no right, then, to abstract sheer impulse from intelligence, to sever two inseparable aspects of one event, and to expect to understand either abstraction isolated from its complementary.

That is to say, we must not hope to be able to give any intelligible account of sheer blind impulse divorced from intelligent foresight.

Here we come upon one of the principal grounds of the general acceptance of mechanistic explanation and the general repudiation of teleological explanation, of the tendency to regard all teleological events as merely mechanism in disguise. We conceive physical things as analysable into ultimate elements each of which retains its causal efficacy, seems to be an intelligible causal agent, an atom or molecule, or an ion, or a particle of electricity, in motion. Hence we conceive mechanistic causation as similarly analysable; we regard any complex physical event as determined by some conjunction of a multitude of distinct causes each operating independently of the rest, each an ultimate causal agency. Teleological causation on the other hand implies complexity of organization. Therefore, we feel that mechanistic causation is fundamental, and teleological causation derived by complication as a special disguised form of mechanistic. But we are beginning to see the flaw in this way of thinking.[1] We are beginning to see that organization pervades nature everywhere; that we cannot validly conceive an atom of any sort going about with all its properties and causal efficacy within it. Everything is bound up in a web of causal relations without which it is not; to conceive it as existing without

[1] This argument for the reality of mechanistic and the unreality of teleological causation has remained hitherto, to the best of my knowledge, implicit; nevertheless it has, I feel sure, been of great influence.

such relations is to misconceive it.[1] Neither a physical atom, nor a sensation, nor a soul, may be validly conceived as a pure and simple substance which takes part in events by *entering into relations* with other self-contained entities.[2] In other words, relations are constitutional or constitutive of things : and organization is universal and primordial, not something superadded, suddenly or gradually, to a world of things that might have existed without it.

If this be sound doctrine (and there seems little room for doubt that it has come to stay) there remains no rational ground for the prejudice in favour of mechanistic explanation as ultimate and final over against teleological as derived and provisional only. We may with impartial minds contemplate the possibilities that both kinds of causation are equally real and ultimate, or that one may be but the appearance of, a disguised form of, the other. And, when we raise the latter question, it is at least as reasonable to suppose that all mechanistic process is teleology in disguise as that all teleological causation is disguised mechanism.

This false way of conceiving instances of mechanistic causation as analysable into a sum of elementary

[1] The most influential exponents of this new way of thinking are, perhaps, Prof. A. N. Whitehead on the one hand, and on the other Prof. W. Köhler and his colleagues of the *Gestalt*, or configuration, school of psychology. The former starts from consideration of physical events ; the latter from facts of human perception and seek to extend the configuration principle to the physical world. The two parties have not yet met ; they do not seem to be aware of their community of doctrine.

[2] It was this false way of conceiving things that gave to Atomic Materialism its specious air of simplicity and finality and rendered it so attractive to many minds.

causes and effects is the chief but not the only ground of the prejudice in favour of mechanistic explanation. In many minds it goes together with, and is supported by, a naïve belief that we can directly observe and fully understand simple instances of mechanical causation. We see one billiard ball impinge upon another ; and we feel the impact and momentum of the ball we catch in the hand. Such communication of momentum from one solid body to another is apt to seem simple, natural and thoroughly intelligible ; and we are ready to accept it as the type of all causation. Of course, with the analysis of the simple atom into a whirling multitude of widely spaced particles of some lesser kind, into a kink or a hole or a vortex in the ether, or into anything other than its simple bullet form, this simplicity and seeming intelligibility of mechanical causation disappears.

The law of the conservation of energy has had a similar influence. It has seemed to be universally true ; and it has been formulated within the framework of the mechanistic scheme. But we must remember that it is made true only by introducing the fiction of potential energy, and that it says nothing about the time of release or conversion of potential to active energy and nothing about the direction of the application of energy. It is only when combined with Atomic Materialism that it takes the form of the law of conservation of momentum and, in that form, rules out all other than mechanical causation.

Another curiosity of so-called scientific thinking sometimes adduced is the assertion that it is the aim of science to predict events ; that only the mechanistic theory enables us to predict and, therefore, it is the only valid or acceptable one. This is to dictate what the nature of things must be in accordance with our desires ; and thus to make our desires the arbiters of reality. This way of thinking has certainly no more validity than the opposite one which points out that if mechanistic causation is universal and exclusive, we cannot hope 'to mould things nearer to the heart's desire' ; and that the assumption must, therefore, be rejected.

We have now defined the peculiarities of teleological causation as it appears in the instances best known to us. And we have examined and found to be without solid basis the prejudice in favour of mechanistic explanations of all events. It remains to examine the principal attempts to explain away teleological causation. This examination must inevitably be long and technical and is, therefore, presented in a note.[1] It must suffice to say here that no one of these attempts seems to be successful or even plausible or promising.

[1] Note 6. *Attempts to exhibit Teleological Causation as crypto-mechanistic.*

CHAPTER IV

MEMORY AND PSYCHICAL RESEARCH

THERE is good reason for believing that some organisms display intelligent purposive activities that owe nothing to the previous experience of the individual organism. But, in all the actions we perform when we are sufficiently developed to study them, prior experience of things and events similar to those with which our action is concerned seems somehow to co-operate in the guidance of action.

This co-operation of the past in the guidance of present action towards the future is what in the broadest sense of the word we call 'memory'. It may seem, and to many minds it has seemed, that the function of memory, being the determination of the course of present action in accordance with past events, is obviously mechanistic; that it lends itself readily to mechanistic explanation and is to be regarded as of the essence of the mechanistic causation of events that seem on inspection to be teleological. For we have seen that the essence of mechanistic explanation is that it proceeds by reference to past events and without reference to the future. This way of looking at the facts of memory is indeed one of the strong points of Modern Materialism and has been much insisted upon by its exponents. Thus the late Jacques Loeb, one of the most thorough-going of modern materialists, proposed to define Mind as the working of associative memory. And the old-fashioned association-psychology, which was dominant throughout the nineteenth century, claimed to give complete explanations of all human action and of all mental activity in terms of associative memory, that is to say, essentially mechanistic explanations. It supported this claim by adducing evidence to show that all memory and therefore, in its view, all mental activity was determined by the material structure of the brain.

Let us examine separately these two claims, that memory works mechanistically and that it is wholly conditioned by material structure. If both are well founded, they together give very strong support to Modern Materialism. If *either* proves to be fallacious, that support falls to the ground.

Let us begin by making what may seem a large admission. Certain authors, confronted with the baffling problems of memory, propose to cut the Gordian Knot in very summary fashion. Mr. B. Russell, for example,[1] disposes of the problem very simply by adopting the adjective 'mnemic' and proposing to recognize a special form of causation called 'mnemic causation': he would by the use of this term merely recognize the fact that past events in the mental life of the individual influence the course of present events, and would leave it at that. This, however, is merely to cloak our ignorance with a Greek word; a popular procedure, but one that lacks finality. We cannot long remain content with a bare statement of fact, no matter how general the

[1] In his *Analysis of Mind*.

statement nor how learned the language in which it is made. If the events of a year ago influence the course of present events, we inevitably and properly seek to fill the gap between them by postulating some changed condition produced by the past event and enduring as an effect that takes part in the causation of the present event. We may ultimately be driven to accept causation at a distance in space; but causation at a distance in both time and space, as implied by the expression 'mnemic causation', is too mysterious to be acceptable save as a last resort.

Prof. Bergson's account of memory is hardly more satisfactory. After brilliantly demonstrating the wide difference between habit and true memory, he tells us that, while the former is the expression of nervous organization determined by previous repetition of actions of the same kind, memory is not to be so explained. But he throws no light upon the facts of memory, beyond stating that the organism carries all its past along with it down the stream of time; which again is merely a highly generalized statement of the facts.

The facts of memory imply that past events in which the organism took part have left upon it some mark, have produced in its organization some effect, that endures and in turn plays a part in determining present action. There is no escape from this inference except by denying time and causation; which would be to abandon all attempt to understand.

As so far stated, then, memory seems to be a phenomenon of mechanistic type. And if, as is true of habit, it involved nothing more than repetition or reproduction of the past, it would remain in that category. But memory plays its part in guiding present action directed towards the future. The future event that we desire to make real is conceived in the light of past experience ; and our choice of means for its realization is largely determined in the same way. It is true that the developed mind may indulge long trains of reverie that seem to be wholly reminiscent. But the primary and fundamental function of memory is to shape our anticipations, to mould our valuations, to influence our desire and aversion, and to render our course of action more direct and efficient in the attainment of its goal. To remember explicitly is the occasional exercise of a highly evolved and late-acquired function ; to anticipate, to look forward, to strive towards a goal however vaguely represented and however near in time, is a primary characteristic of all intelligent purposive action, that is to say of all mental activities, both those in which the individual's past history seems to play no essential determinative role and those in which memory is a prominent factor.

Further, it is to be noted that, even where the memory function predominates, as in sheer reminiscence, it is controlled by purpose and interest or, broadly, by conation, by interested or purposive striving. We retain in memory that which interests us, that which is relevant to our purposes, our strivings towards goals. And, when we ransack the stores of memory, that which is relevant to the dominant purpose of the moment comes back to consciousness ; recollection is highly selective ;

and the selection is governed by conation. If it were not thus governed, if it were merely a mechanical reproduction of past sequences according to the laws of association, it would be of little help to us in guiding present action.

Memory, then, though it is determined by the past, works towards the future *as one aspect of an essentially teleological or goal-seeking activity*. And, in thus working to mould the future as we conceive and desire it, and to guide our forward striving action, it becomes imagination, the function in which Mind manifests most clearly its creative power.

THE BASIS OF MEMORY

Turn now to the other great problem of memory. What is the nature of those traces of past activity which we bear with us for the shaping of present striving towards the future? They are parts of our organization, of the organized structure of the individual. This problem is therefore part of a still larger problem, namely: Is our organization wholly material in the same sense as that of a machine? Is it wholly spread out in three-dimensional space and perceptible in all its parts by our senses? Does its integrity of function depend wholly upon the maintenance of the spatial relations of the parts within narrow limits of variation?

This is a question to be answered only in the light of the empirical evidence; and much evidence can be adduced in support of either answer—all of it indirect.

Biological Materialism (with the great majority of

biologists) says : Yes, the organization of the living creature is wholly material ; it is a distribution in space of chemical substances which in the processes of metabolism illustrate only the laws that are revealed by the study of inorganic processes. The confidence with which the truth of this view is assumed is in large part a hang-over from the palmy days of Atomic Materialism, the days when we were challenged to imagine two molecules bound together by the thought of a beefsteak. The purely material view of organization could be deduced with confidence from Atomic Materialism. But, although it accords well with Modern Materialism and, if it could be independently established, would lend support to it, it cannot be deduced from it ; and no substantial foundation for it seems to exist. The very general acceptance of it seems to be due to that peculiarity of our minds on which Bergson has so strongly insisted, namely, our preference for thinking in geometrical terms. That peculiarity again is one that very naturally results from the fact that the survival of the individual and of the race has required through long ages effective thinking of this sort for dealing with the spatial relations of things. We cannot, however, safely deduce from the natural bent or bias of our minds any conclusion as to the nature of things, not even the conclusion that we have no minds.

In addition to this vague general prejudice, two lines of empirical evidence are mainly relied on by biologists as support for the purely material view of vital organization. First, it is said, we never observe any evidence of vital organization other than the development and movements of material organisms. Secondly, there is always material continuity between one state of organization and another; and if this material continuity is broken, organization ceases to be manifested; and if it is forcibly disturbed or deranged, the manifestations of organization are deranged.

In criticism of this argument we may, first, question whether, if the premises be granted, the conclusion follows; secondly, we may inquire whether the premises are well established as inductive generalizations.

Suppose, then, it be true that we never observe any manifestation of vital organization but in material structures or organisms. Clearly, that fact, if fact it be, does not in itself show, or allow us to conclude, that all that is essential in organization is the spatial arrangement of matter. We are so constituted that we tend to conceive all sense-impressions as excited by material objects. Yet certain doubtful cases should give us pause. We observe a wonderful display of the Northern Light, or some effect of an electric current passed through some part of the body. Or we become aware of the pain or the intention of a friend, or of the fury of a mob or of an animal, and we shape our actions in accordance with our anticipation of the further course of events.[1]

[1] I remind the reader of the discussion of perception where we noted the falsity of the view that we perceive only the sense-qualities of things and the fact that, just as we perceive their causal efficacy, so also we perceive in the face or the voice of a friend his intention, his emotion, his disappointment or satisfaction.

In instances of the former kind the physicists give us a large variety of theories to choose from, most of them involving reference to particles or quanta or ions or charges of electricity, or spatial distributions of energy. But, though the Aurora at least appears spread out in space, it is by no means made clear that we have to do with a material organization. In instances of the latter class we have no ground whatever for the view that a pain, a desire, an intention or an emotion is in any sense a material fact or a spatially distributed event. Our actual sense-impressions are, as we have seen, but symbols that prompt us to think the nature of the objects we perceive. And the only test of the degree of the truth or validity of the way we conceive the objects is the pragmatic one, the success of actions guided by our thinking. The argument attempts to prove a negative and, in this notoriously difficult task, it fails. It attempts to prove that in vital organization there is nothing that counts but matter and its spatial arrangement.

The second part of the argument we are examining is no more conclusive. Its premises are that there is always material continuity of vital organization, and that derangement of the material structure always results in disturbance (amounting in extreme cases to abolition) of all evidences of organization. From these premises the conclusion is drawn that the material collocation is all that counts in the organization.

There are two main lines of empirical evidence, both of which go far to support the premises ; but at the same time they refute the conclusion of the argument. The premises are that function is correlated with *material* structure. Now, if it could be shown that this correlation is perfect (in technical terms, that the coefficient of correlation is $+ 1$) then, and only then, would the conclusion follow, namely, that function is wholly determined by material structure.

The one line is the evidence afforded by study of the development of the egg, the processes of ontogenesis, of heredity. The egg or germ-cell seems to be an essential material link between parent and offspring. And modern research has shown that certain particles of the germ are correlated with certain organs and functions of the organism that develops from it, and justifies the view that this is true of all parts and functions. But, if this be granted, the conclusion does not follow. In order to establish the conclusion it would be necessary to show that all the organs and functions of the developing organism are perfectly correlated with the peculiar spatial arrangement of the particles of the germ.

Now experimental observation does not reveal any such complete correlation. On the contrary, it proves that the correlation is very partial only, and that, therefore, the organization of the germ must comprise something over and above its material structure. The fact of principal importance here is abundantly established and undisputed ; it is that the spatial arrangements of the material constituents of the germ (or of the developing organism) may be grossly distorted, and, nevertheless, the course of development may be rectified, may return to the normal and result in a normally constructed and functioning organism.[1]

The second line of empirical evidence of special importance is the effect on mental life, more especially on memory, of disturbance of the structure of the brain. This is a field in which an immense number of observations have been gathered, experimental and other.

In the nineteenth century were made many observations that seemed to point to an exact localization in the brain of all mental functions. And, if the anticipation of the completeness of this correlation, then widely entertained, had proved well founded, such proof would have been strong evidence of the entire dependence of function on material structure. But further research has shown more and more clearly that this anticipation did not correspond to the facts; that the correlation between details of brain structure and details of mental processes cannot be carried beyond a very restricted field. As was to be expected on any

[1] Especially Hans Driesch's work reported in his *Science and Philosophy of the Organism*. The facts of regeneration of ablated parts of a completed organism are a special case of this general truth. For example, a limb or part of a limb of a newt may be cut off at any plane; and the missing part is regenerated exactly fitting to the surface of ablation. Is this a function of, is it perfectly correlated with, any material collocation? Certainly we cannot ascribe it to the cells of the surface of ablation alone. We can only say that it is a function of the whole organism. But then we can lop off another limb without disturbing the process of regeneration of the former. Therefore it is not a function of, is not correlated with, the spatial collocations of the whole of the organism. And there would seem to be no possibility of perfectly correlating it with the material collocation of any one part.

informed view of the possibilities, there seems to be close correlation between, on the one hand, experiences of the various sense-qualities and of various bodily movements and, on the other hand, certain parts of the brain structure. But all attempts to carry the correlation farther have broken down.

Prof. Bergson [1] has carefully examined a mass of the empirical evidence and shown clearly this lack of detailed correlation. And since his review was made further evidence of the same kind has accumulated.[2] In general, it may be said, the evidence supports the view that in some sense the brain functions as a whole ; and that, when one part is destroyed, other parts can in a surprising manner take over as it were the impaired functions ; or, at least, the impaired functions are restored and become correlated with parts of the brain other than those with which they were originally or normally correlated. These facts are even more opposed to the view that function depends altogether on material structure than are the instances of

[1] In his *Matter and Memory*.

[2] I will refer here only to the conclusions recently announced by Dr. J. K. Lashley. They are the more weighty since Lashley seems to have a strong mechanistic bias and would be shocked to know of the application of his conclusions which I am here making. His observations were made by training rats to run mazes before and after ablation of small or large portions of the cerebral hemispheres. The significant result from the point of view of our present problem was that, though ablation of parts of the brain produced in various degrees impairment of this form of intelligent action, there was no correlation found between the memory displayed and any particular parts of the brain, and even ablations of almost the whole of the two cerebral hemispheres did not wholly abolish either the power of learning or the retention of that which had been learnt.

regeneration of bodily organs and functions. For in these cerebral instances the destroyed part of the structure is not regenerated.

Shall we then, like Mr. B. Russell, give up the problem and content ourselves with speaking of 'mnemic causation' as the explanation both of development according to the hereditary pattern and of the facts of memory? No! We must continue to believe that these functions express organization; but we must believe that organization is not completely expressed in material structure, that its nature is such that it cannot be directly revealed to our senses in the form of material collocations.

It is significant that the two lines of evidence which most unmistakably negate the view that vital organization consists wholly in the collocation of matter really converge and merge. Many thinkers have pointed out the analogies between the phenomena of memory and those of heredity; and we can hardly doubt that those are right who go farther, who say that it is more than an analogy that we discover, namely, a true identity of nature.[1] As we have seen, all that we call memory from the psychological view-point is an expression of the organization of the individual as modified by its own activities in the past; and the processes of development (of ontogenesis) are expressions of organization built up by the activities of the ancestors throughout the ages of evolution.[2]

[1] The list is long; I mention only Claude Bernard, E. Hering, Cope, Nägeli, Samuel Butler, James Ward, T. H. Ribot, R. Semon, E. Rignano.

[2] This remains broadly true even if we accept the neo-

Biologists are very ready to admit that memory and heredity are of essentially the same nature, so long as this view is couched in terms of Biological Materialism or, at least, in terms that do not openly challenge that dogma. Thus the lecture given by the physiologist, Ewald Hering, some fifty years ago under the title, 'On Memory as a General Function of Organized Matter', the works of Samuel Butler and of R. Semon,[1] all insist upon this essential identity, all avoid challenging Biological Materialism, and all have been widely acclaimed by biologists. Now the view here taken is the same, except that it goes farther; it does challenge Biological Materialism. While regarding memory as a general function of organized matter; it lays much stress on the organization and less on the matter.

Biological Materialism says that living things and inorganic matter are composed of the same stuff and that the difference between them consists only in the more complex and special spatial arrangements of organized matter. It is here maintained that the empirical evidence is not consistent with the view that the organization of living things consists merely in special spatial arrangements of matter; for there are many instances (of which the most indisputable have been briefly indicated in the foregoing pages) in which function is not closely correlated with

Darwinian or purely selectionist theory of organic evolution. For the various forms of selection produce new species only in so far as the organisms actively struggle for existence. This is less obviously true of plants than of animals; yet the evolution of plants by selection requires at least the perpetual production of an excess of individuals.

[1] *Die Mneme* (Leipzig, 1908) and others.

material structure, but shows rather a relative independence of it.

The gist of the argument that refutes Biological Materialism has now been stated. It is unnecessary to elaborate it or to array in detail the facts on which it rests. But it is worth while to supplement it by an argument of an impressive though less conclusive kind; namely, the difficulty, hitherto unyielding to all attacks, of conceiving any material structure that would constitute an organization capable of expressing itself in the ways in which vital organization expresses itself in the normal course of events; that is to say, quite apart from the excceptional instances of regeneration of parts and of restitution of functions after distortion of the material structure, the normal processes of morphogenesis and self-regulation of living bodies cannot be explained in terms of material structure alone.

This supplementary argument may also be presented under the same two heads, ontogenesis and memory.

Consider the ontogenesis of a particular moth from the egg. The egg is a minute speck of protoplasm containing in a semi-fluid matrix a number of distinguishable microscopic particles or genes arranged in a fairly definite pattern. And no doubt each of these particles may have a definite material structure or pattern of arrangement of its constituent elements, whatever these may be. And there is good evidence that each gene is somehow necessary to the development of one or more of the features of the adult organism. The egg-cell divides again and again; masses of cells become differentiated to form special organs, each vastly complex and exactly fitted to play its part in the economy of the whole. And the whole assemblage of organs, which is the caterpillar, displays in a number of highly complex actions a number of complex instincts. Then a process of resolution sets in. The organs are resolved into a mass of cells within a bag of skin, a mass that seems little more than a drop of pus. Then a new organizing process begins. New organs gradually form, all different from those of the caterpillar, many of them entirely new and more complex; and then the moth emerges to display a new and entirely different array of complex instincts.

Turn to an instance of memory. You, a polyglot in a company of polyglots, are informed, directly or indirectly, of a plot to assassinate an important personage. You find an opportunity to lay your information before the guardians of public order. You remember very positively the substance of the information gathered, the total meaning of the words heard; but you cannot perhaps remember a single one of the words; you may not even remember in what language the information was conveyed. That instance illustrates clearly a general truth, namely, that in remembering facts of perception we do not merely or chiefly or at all reproduce certain sense-impressions made upon us; rather we remember the meaning we discovered in or through or by aid of some complex of sense-impressions; sometimes we can and do reproduce or represent the complex of sense-impressions more or less faithfully, but that is unusual and of altogether secondary importance.

Now all attempts to explain memory as conditioned by material structure have been attempts, by the aid of physical analogies, to show how the material structure in the brain might, like the gramophone plate, reproduce or give back a complex of impressions similar to those made upon it.[1] Such attempts have a certain plausibility; and we cannot say that they are in principle hopeless. But the problem is indefinitely more difficult when we understand its true nature. All the evidence goes to show that, in so far as we can reproduce complex sense-impressions, it is only in so far as we remember the meanings we discovered in them, the objects or realities which they symbolized and the complex relations between them which alone make them interesting to us.

When we keep in view this, the most significant, aspect of memory, it is obvious that no attempt hitherto made to explain the fact in terms of material structure alone has the slightest plausibility; just as no attempt hitherto made to explain ontogenesis in terms of an organization that is purely material has any plausibility. We therefore may find in this failure support for the conclusion we have drawn from the facts which show that such explanation is

[1] This is true of the two most thoroughgoing attempts of the kind hitherto made, that of R. Semon and that of E. Rignano. The latter, by far the most thorough and impressive hitherto made, is examined in Note 7. The vast majority of biological materialists, who confidently assume a purely material basis of memory, do not conceive in the vaguest way what the nature of that basis may be. They are like the savage who believes unshakably in the efficacy of his magical rites, but has no faintest notion of the way in which they may produce the effects he attributes to them.

in principle impossible, the conclusion, namely, that the organization which expresses itself in memory and in ontogenesis, in individual and in racial memory, is not a purely material structure, does not wholly consist in the collocation of chemical substances.

In the light of the foregoing discussion we may consider some of the inanimate things that have been likened to the living organism. The kind of thing that most readily suggests itself is a man-made machine. A machine is like an organism in that each part has a function, each part contributes towards the production of the result for the sake of which the machine exists, and all the parts co-operate harmoniously. But when we have said this we have said all. The machine differs profoundly in all-important respects. It does not grow ; it is put together. If its parts are deranged, it cannot rectify its working. If any part is destroyed, it cannot restore it. It is teleological only in respect of its construction ; not at all in respect of its working.

The flame of a candle is analogous in certain respects. It is a material structure which, in spite of variations of form, constantly returns to its normal form ; it seems to strive to maintain and restore a certain equilibrium. Its matter is perpetually used up and renewed. Further, under appropriate material conditions it can give rise to other flames like itself. It depends upon a constant supply of oxygen, as does the living organism. But the flame has no intrinsic coherence or stability. It lacks just that which the machine possesses. Hence, while in a very broad sense the machine may be said to exhibit memory, in that its present working may be modified by influences brought to bear on it in the past or even by its own workings in the past (may even be improved by them), the flame has no memory. The living organism combines in a high degree the incompatible properties of the machine and of the flame. Like the flame, its matter is in constant flux and change, yet preserves and restores its form ; like the machine, it has internal continuity and intrinsic stability and bears its past history with it. If we could construct a machine of flame alone, we should have something more analogous to the organism. But we cannot ; for they have incompatible properties.

A helpful analogy is the following. Imagine a shower of sparks, glowing particles of iron, projected from a nozzle within a magnetic field. They present a certain pattern to our perception, one which in many respects would resemble a flame. The Biological Materialist would explain the pattern from the properties of the particles, the mass, the weight, the momentum, the mutual attractions and repulsions of the particles. But such an explanation could never be complete, because it leaves out of its account the magnetic field with its peculiar pattern.[1] Just so, the organism and its life processes cannot be explained by taking into account only the properties and collocations of its material particles ; there is in its organization something else, something

[1] I am assuming for the purpose of the illustration that the particles retain their magnetic properties at high temperature.

analogous to the magnetic field, that plays an essential part in the maintenance of form and function, modifying profoundly the reciprocal influences exerted by the material particles upon one another. Cut off the flow of sparks and the pattern disappears; but the magnetic field remains. Just so, perhaps, when the matter of the body is dissipated, there may remain unseen that pattern of forces which makes of the collocation of material particles a living organism and is of the essence of its organization.

Or perhaps a variation of this picture would offer a truer analogy. Suppose that each of the glowing particles is itself a magnet. Then the pattern presented by the shower of sparks will owe its nature, its peculiar modes of behaviour, in large measure to the reciprocal magnetic influences of the particles. The Biological Materialist is like a physicist, who, knowing nothing of magnetism, should persist in attempting to explain all the phenomena presented by the shower in terms of those properties only that are known to him.[1]

We get a yet truer analogy if we combine the two pictures. The particles projected into the magnetic field become magnets under its influence; and their behaviour is then in large measure a function of the influence of the field and of their reciprocal magnetic influences. In a similar way the matter

[1] This, be it noted, was the actual procedure of physicists under the dominance of Atomic Materialism. They ignored, in their attempts to explain, the many imperceptible factors, forces, modes of influence, or forms of energy which the modern physicist has been driven to infer in the course of his efforts to explain physical phenomena more fully.

that is assimilated into the body of a living organism becomes profoundly modified as it enters into the relations that are peculiar to living things; in the living body it is brought under influences that are not to be found elsewhere; and, under those influences, it takes part in events or phenomena of a peculiar nature, namely, teleological events. The fact that these peculiar factors or influences do not directly evoke in us corresponding sense-qualities is no more a ground for denying their reality than in the case of such factors of purely physical events as magnetic fields, electric strain, chemical attraction or repulsion.[1]

Life is a continuing activity. It may be that it is manifested to us only when matter exists in the peculiar collocation we call protoplasm; just as magnetism is manifested only where certain unusual collocations of matter are formed. And as we cannot understand the phenomena presented by collocations of the latter kind in a magnetic field without taking account of the peculiar nature of that imperceptible field, so we cannot hope to understand the phenomena presented by living protoplasm without taking account of the imperceptible factor that makes the difference between protoplasm living and protoplasm dead.

The phenomena presented by magnetic particles in a magnetic field cannot be explained from the properties of those particles alone; the field is a factor which may have existed before they came under its influence and may continue to exist when

[1] Cp. Note 8. *The Legitimacy of postulating Non-spatial Factors in Vital Organization.*

they are withdrawn from it. In a similar way the life-factor may be all pervasive ; yet manifested to us only through peculiar collocations of matter. But the parallel is not complete ; we are acquainted with Life directly ; each of us is directly aware of his own life-activities, has some direct acquaintance with his own strivings. Since, then, to each of us, life, his own life, is directly revealed without the mediation of sense-impressions, it may be that other life also may be manifested to us in some more direct fashion than through sense-impressions. And of this possibility there is some evidence, the more important part of which we must briefly consider.

THE ASSUMPTIONS OF BIOLOGICAL MATERIALISM ARE QUESTIONABLE

Hitherto we have granted the premises of the argument of Biological Materialism, namely, that all expressions of vital organization are through the medium of matter or are directly associated with material structures. But we are now to review certain empirical grounds for questioning this premise.

We enter here a field of alleged facts which have never been acknowledged as proper objects of scientific investigation and towards which most men of science (just because the alleged facts are irreconcilable with the postulate that all organization is purely material) have exhibited a dogmatic incapacity to weigh evidence—the field set apart for curious amateurs under the name ' psychical research '.[1]

[1] It is the biologists rather than the physicists who have displayed a positive hostility towards even the investigation of

I am not going to trouble you with any of the much disputed evidence for so-called super-normal physical phenomena ; for the good reason that the evidence for any such phenomena seems to me, after giving some attention to the question during many years, extremely unsatisfactory and questionable.

I ask attention to alleged facts of two classes only ; namely, those grouped under the head of telepathy, experimental or spontaneous and sporadic ; and those which seem to imply that human personality can and does in some instances continue to manifest itself, manifest something of its peculiar organization, its memory, after destruction of the bodily structure.

It is impossible to present here, even in the most summary fashion, the evidence of telepathy. I must be content to say that evidence of it has been carefully and critically gathered [1] through a period of some fifty years ; and that during this period the various kinds of evidence have continued to grow in bulk and weight, until now it seems irresistible by any competent person who may consider it comprehensively and impartially. The only way of escaping the conclusion to which it points is to assume that all, or the great majority of, those persons who have laboured to gather the evidence are incompetent and biased cranks. But this assumption can be made only by those who have no

such alleged facts ; for the postulate of merely material organization with which they conflict is a dogma of Biological Materialism, not of physics.

[1] Chiefly by the English Society for Psychical Research.

acquaintance with the persons concerned and who neglect to consider their status, their reputation, and their achievements in other lines of activity.

The conclusion to which the great mass of concordant evidence so strongly points is that, under conditions which we have not yet been able to define satisfactorily, the mental activity of one person may influence that of another in the direction of assimilating it to his own, without the demonstrable mediation of any physical process, without the transmission through the intervening space of any physical process of any kind known to science. Many of those who are aware of the strength of the evidence for telepathy have inclined to believe that all instances of it imply some hitherto unknown kind of physical mediation between the two persons concerned. This, of course, is a possibility that deserves to be thoroughly explored. But many of the facts are strongly opposed to it, if not quite incompatible with it. And in view of the fact that we all accept the reality of gravitational attraction without having any knowledge of any physical process by means of which it may be transmitted, insistence upon the demand for physical mediation seems anomalous. It is true we have the ether; but we have no satisfactory suggestion as to how the ether transmits or mediates gravitation. In the case of gravity, the transmission process may be conceived to be of a relatively simple uniform type; whereas a telepathic transmission process would have to be conceived as highly complex and susceptible of infinite variation; and this fact makes more difficult the assumption of physical mediation in the latter case. But what makes the assumption of a physical mediating process most difficult is the fact that the telepathic effect seems to be relatively independent of distance in space and to depend mainly upon such distinctively mental conditions as intensity of interest or emotional congruity of the persons concerned.

We seem, then, to have in telepathy a mode of manifestation to one person of the life-activities of another, a mode which is an exception to the generalization that the life of other beings is manifested to each one of us only through physical media and impressions on the senses.

SURVIVAL OF PERSONALITY

Let us now consider equally briefly the empirical evidence that personality or some part of personality may survive the destruction of the body.[1] The gathering and the evaluation of this evidence is a task of very extraordinary delicacy and difficulty. This task also has been prosecuted during half a

[1] As anyone may see on running through the series of ' Ingersoll Lectures ' delivered annually for many years by distinguished philosophers, modern philosophy shows a strong tendency to burke this question, to sit on the fence, or to assure us of immortality or survival in some sense that has neither interest nor significance for the plain man ; in short, in some purely Pickwickian sense. This is inevitable in so far as the question raised is the question of fact rather than the question of value. But it may fairly be complained that, with very few exceptions, the Ingersoll lecturers assert the value of survival, and, as regards the question of fact, seek to put us off with answers that are no answers

century with great care by a small group of competent persons ; and the results attained are neither negative nor negligible. They may best be summarized in the assertion that they place the unbiased inquirer before a dilemma. Either personality is not in all cases utterly dissolved with the destruction of the body, or telepathic communication of a most far-reaching and improbable kind occurs.[1]

If we accept the latter alternative, the evidence for telepathy is strengthened, and especially that for telepathic communication without physical mediation.

Either alternative, then, provides good ground for belief in events of a type in which matter plays no essential part. But, since the telepathic alternative seems difficult in itself as an explanation of the phenomena in question, by reason of the immense range and peculiar and purposeless selectivity which would be implied, and since it does not cover all of the facts, we may consider what is involved in the acceptance of the other alternative, the survival of something of personality after bodily death. I say 'something of personality'; for impartial consideration leads inevitably to the view that whatever of personality may survive must be in many respects different from the personality that was manifested in the flesh. And this, of course, is the popular, the orthodox, perhaps one might say, the theological view.

The empirical evidence is not at present of such a

[1] It is chiefly the possibility of such communication that renders the evaluation of the evidence for survival so remarkably difficult.

nature as would justify any attempt to say how much of personality survives. It does imply that something of what we may call the higher aspects of personality may survive, especially intimate personal memories and interests. Now the difficulty and, from the point of view of the problem of these lectures, the chief interest in accepting the conclusion to which such evidence points are that it implies survival of some part of the total organization that we call personality.

As will be shown in our next lecture, many men of science are now prepared to admit the causal efficacy of our mental activity. The words 'emergence' and 'emergent evolution' have worked wonderfully, one might almost say magically, in this matter; with the consequence that to-day a man of science may make such admission openly and find himself in good company; whereas an inclination to such admission shown a decade or two ago sufficed to render him an outcast.

But, though it has now become scientifically respectable to admit the causal efficacy of mental process, it is still disreputable to postulate any enduring organization that is not material. Is this prevalent objection well-founded? Or is it merely a prejudice surviving from the age of Atomic Materialism? I incline strongly to the second view, and find support in the following considerations. The distinction between substance and process has long had a prominent place in philosophy; but many modern philosophers have rejected it. And we see the same tendency in modern science. We see it in the recent breakdown of the distinction between matter and energy. We see it expressed by those who, like Whitehead, would describe all nature as consisting of events rather than of things. We see it in the dictum of Alexander that all things are motions.[1] We see the same tendency carried to a sort of half-way position in certain recent proposals to postulate a kind of organization that is physical and yet not material. The best instance of this kind known to me is embodied in the little book by a distinguished mathematician to which I have already made reference.[2] He writes: 'If it appears that the theory of engrams in protoplasmic structure fails to account for the psychic transformations of memory, we must examine the alternative hypothesis of traces in psychic structure. To surmount the difficulty we have to abandon our natural prejudice against recognizing something which our senses do not reveal to us. But we must remember that, except in the case of light and heat, our sense organs will only sense matter. Only a restrained scientific imagination which takes the form of a hypothesis can carry us farther. In the case of the physical sciences we are up against an impassable boundary, unless we can envisage the etherial vibrations which enter into nearly every branch of physics. And in the case of the biological sciences we are also up against an impassable boundary unless we can envisage the impalpable psychical structure which enters into all mental phenomena.'

[1] *Space, Time and Deity* (London, 1920).

[2] *The Basis of Memory* by W. R. Bousfield, F.R.S. The same view is implied in the larger and earlier work by the same author, *The Mind and its Mechanism* (London, 1926).

Bousfield proceeds to formulate an 'alternative hypothesis, according to which the basis of memory is not a record in protoplasmic structure but in an immaterial " psychical structure ". Such a structure involves the conception of a substance of which it is built. We have seen that there are many phenomena which the hypothesis of engrams in protoplasm will not explain. Yet engrams or mnemonic traces of a sort there must be in some substance, and this substance we shall call psychoplasm, a substance which we postulate as consisting not of material protons and electrons but of some other modification of the ether. This substance, which we call psychoplasm is as hypothetical as the ether, and no more so. Electrons and protons are etherial, but material, and do not satisfy the required conditions. Psychoplasm we may regard as a modification of the ether, and therefore as physical though immaterial. Our definition of matter, according to the present state of science, is a substance consisting of protons and electrons. As the basis of psychoplasm, we postulate some modification of the ether which is other than protons and electrons and therefore does not come under the definition of matter. . . . Psychoplasm in our vocabulary stands for the immaterial substance of which the " psychical structure " is built.'[1] Further,

[1] The author is avowedly attempting to render more concrete and precise the mental or psychical structure postulated in my *Body and Mind* and in other of my books. It may be of interest to readers of those books to read the following extracts from *The Basis of Memory*: 'The hypothesis presented differs somewhat from McDougall's conception since it contemplates not a metaphorical, but a real and physical structure (for every

'Our hypothesis is that every living cell has not only a protoplasmic structure, which is the material basis of life, but also a psychoplasmic structure, which is the basis of psychic life, capable of being affected by psychic factors and of retaining mnemonic traces.'

Bousfield nowhere states that he conceives the psychoplasmic structure as extended in space ; but, since ether is conceived for the purpose of accounting for transmission of energy across wide spaces, and is generally described as filling space, we may fairly assume that psychoplasm is to be conceived as extended in space.

The same may be said of an older proposal related to Bousfield's, namely, W. Ostwald's proposal to overcome materialism by abolishing matter entirely from the conceptual armamentarium of science and to replace it by an array of energies of various kinds, among which he proposed to recognize a psychic energy.[1]

I find myself very sympathetic to Bousfield's proposal. Yet I am inclined to prefer my own, namely, a psychic or mental structure that is not extended in space, but may rather be described, in

etherial derivative must be regarded as physical) which co-operates with the material brain and serves as a link between sensorial impressions and consciousness. Furthermore, the hypothesis includes in its purview the slightly different concept of " mental structure " which is to be found in McDougall's Essay on Mental Evolution, in which he refers to the *innate* or *racial mental structure* which is the carrier of instincts, and to the " *complex mental structure* " which is built up by experience. And it should be noted that it links up with McDougall's original conception of an immaterial organization which serves for the integration of sensory stimuli.'

[1] *Vorlesumgen u. Naturphilosophie* (Leipzig, 1903).

Prof. Driesch's terms, as a qualitative manifold which, while not spatial, yet acts into space. And the ground of my preference is as follows. The psychic structure that I conceive is called by Bousfield 'metaphorical'; the implication being that whatever is not physical and spatial is metaphorical; and I take it that, as thus used, metaphorical means unreal.

Now, ever since Kant pointed out that perhaps we conceive the physical world as spatially extended merely because it is the nature of our minds so to conceive it, there have been three parties among philosophers: those who conceive everything as really extended; those who regard space as merely a mode under which we conceive all reality; and thirdly those who, while recognizing that our minds tend strongly to conceive all things as spatial, believe that this way of thinking may be true or valid of the physical world but not necessarily true of all things.[1] I have never encountered any demonstration of the falsity of this third view. The common rejection of it by philosophers seems to be due to the common demand for a monistic account of the universe, the demand that all things shall be of one kind and fit neatly into one picture. This demand in turn seems to be justified only on aesthetic grounds, the ground that such a picture of the universe would be more satisfying aesthetically than any one not conforming to the prescription. It is in short a clear instance of that kind of philosophic error which I deprecated in the first lecture,

[1] Professors Bergson and Driesch belong to this third group, as also, I suppose, all the neo-scholastics.

namely, the attempt to dictate the nature of things in accordance with our scheme of values. The reasoning when explicated runs, briefly, as follows : We should like the universe to be all of one piece and texture, therefore it must be so.[1]

There is, then, no sufficient *a priori* ground for assuming that what is non-spatial is unreal, or that if anything is spatial everything must be spatial. It may be that our spatial way of conceiving things is more nearly true, is relatively more valid, of physical things and events, and less valid or quite invalid, i.e. obstructive rather than helpful, when applied to mental events. And there are two good grounds for preferring to conceive our organization as in part non-spatial, as comprising a non-spatial factor. First, we have positive knowledge of and immediate acquaintance with instances of organization that are non-spatial : for example, the organization of a sonata, the aesthetic structure of a drama or poem, the logical structure or organization of an argument. And let it be noted that, in all such instances, the non-spatial structure or organization of the whole governs the actual course of serial presentation of the whole as a temporal and spatial manifestation.

Secondly, we have found good reason to believe that teleological causation is real. Now such non-spatial organizations as those referred to in the foregoing paragraph are teleologically engendered

[1] Modern philosophers seldom pause to justify or to state the grounds of this preference ; they are apt to accept it as something that goes without saying, and to regard the demonstration that any view does not comport with their preference as a *reductio ad absurdum*.

and, when they govern the spatial and temporal presentation of their structure, they operate teleologically. It would seem to be a true empirical generalization to say that, so far as we can observe, all spatial organizations work mechanistically, and all non-spatial organizations work teleologically. And this generalization affords at least a presumption in favour of the view that any organization that works teleologically is non-spatial.[1]

Our brief review of the principal facts of organization points, then, to a conclusion concordant with our review of teleological causation ; namely, that, just as the actions of the living organism cannot be explained, nor even intelligibly described, in purely mechanistic terms, so also its organization cannot be completely described in terms of material structure. The facts of both orders combine in pointing to non-spatial organization that expresses itself with a causal efficacy that is teleological.

It is perhaps worth while to point out how the conclusion reached in this chapter differs from the teachings of other vitalists. The essence of vitalism in the widest sense consists in the denial that the laws expressed in the events of the physical realm are adequate to the explanation of vital events.

[1] The teleological nature of mentally directed action which is the foundation of this reasoning has been neglected by Bousfield, as also by Ostwald when proposing to recognize a psychic form of energy. It is recognized by Rignano in his revival of Ostwald's suggestion of a special form of energy to be called psychic or mental. But we cannot accept an intrinsically teleological form of energy, because, as we have seen, teleological action implies organization, whereas an energy of any kind is not in itself organized. Cp. Note 8.

Two forms of vitalism are usefully distinguished, the emergent vitalism of Emergent Evolution (q.v. Chapter V) and substantial vitalism. Substantial vitalism differs from emergent vitalism in postulating in living beings some kind of causally efficacious reality that is not operative in purely physical events. The older form of vitalism that postulated merely some vaguely conceived 'vital force', and the vitalisms of Bergson and Driesch, are of the substantial variety. The vitalism here indicated is of the same variety, but differs from these others in regarding the non-physical non-mechanical factor in vital activity as an organized part of the organism, rather than as something that acts upon the organism from outside it.[1] The implication of this something extrinsic to the organism, yet acting upon it to modify the physico-chemical events within it, is, I think, one of the features of the more usual forms of substantial vitalism to which strong objection is widely felt.

[1] I do not feel sure that I am right in saying that Prof. Driesch's entelechy is regarded by him as extrinsic to the organism.

CHAPTER V

EMERGENT EVOLUTION

IN the late decades of the nineteenth century when Atomic Materialism was still in vogue and when Darwin and Spencer had given general currency to the theory of evolution, both organic and inorganic, the triumph of the neo-Darwinian school led by Weismann seemed to complete the edifice of Modern Materialism. For neo-Darwinism claimed to show that the purely mechanistic processes of natural selection and fortuitous variation had sufficed to bring into existence all species of plants and animals with all their marvellous adaptations and powers of new adaptation. Teleological causation, intelligent purposive action, was ruled out of the scheme; for, if the claim was well founded, all such causation was otiose; the scheme of evolutionary explanation was complete without it. Mind became merely 'consciousness', a somewhat mysterious by-product, a mere train of passive reflections of things and events; which, like the physical reflections in a mirror, have no influence upon the events of which they are in some sense copies.

And yet, there was something queer about this scheme. Even the reflections in a mirror are not produced without expenditure of energy; and the reflections produce further consequences, which, in some cases, may be important. Those biologists who stood fast on Biological Materialism, and who were not indifferent to these difficulties of Epiphenomenalism, preferred therefore the theory of Psycho-physical Parallelism, according to which the stream of consciousness is not in any sense produced by the physical processes of the body, has no causal relations with them, but merely runs parallel with them, as one railway line may be said to run parallel to its fellow, following all its twistings and turnings, but never interacting with it. As a descriptive formula this had much plausibility; for it was not easy to point to any clear exception. And it had two great advantages: it left the scheme of mechanistic causation intact; and it pooled all the obscurities of the relation of mind to body in one absolute mystery, the mystery of the alleged parallelism of the two streams of events without interaction between them. Yet it was an unsatisfactory formula: Science cannot long rest content before an utterly mysterious but obtrusive correlation. Hence, although in the opening years of this century Psycho-physical Parallelism was the orthodox creed of psychologists and biologists, it has now disappeared from the scene, or survives merely as an historical curiosity still hugged by a few intellectual tortoises.

In a series of articles published in the year 1897,[1] I ventured to suggest that we may reasonably believe certain brain-processes (those of the synaptic junctions) to engender sensory affections or qualities of consciousness; that these become synthesized in complex wholes according to peculiar

[1] 'A Contribution to an Improvement of Psychological Method' (*Mind*, N.S., vol. VII).

laws of synthetic combination ; and that these in turn react upon the brain processes modifying the course of those events.

This was a well-meant attempt to rescue Mind from the undignified position to which it had been reduced by the prevalence of Epiphenomenalism and Psycho-physical Parallelism. It did not seem to me very satisfactory, even at that remote date. Yet, strange to say, this view, or something very like it, seems now in a fair way to become orthodox doctrine. This change has been due, not in the least to my advocacy, but rather to the magic of a word, a word to which new meaning has been attached, the word ' emergence ', together with the expression ' emergent evolution ', and the German word *Gestalt* with its English equivalent ' configuration '.

There can be no question that this movement of modern thought of which the word ' emergence ' is the banner, is a very remarkable and important one. Many thinkers of very different schools are converging towards the one centre ; some of them seemingly ignorant of those who are marching along other of the convergent lines.

The leaders of the *Gestalt* school have shown no interest in Emergent Evolution and are apparently largely concerned to save the principle of Psycho-physical Parallelism. Professors Alexander and Lloyd Morgan are concerned chiefly to save God and the coherence of the evolutionary scheme. Professor Whitehead is concerned to save the unity of the natural world and seems indifferent and even a little hostile to the *Gestalt* workers, and not directly interested in biological evolution or the psycho-physical problem. Messrs. Noble and Smuts are free lances, each concerned to secure a hearing for his own peculiar thesis, and showing little evidence of influence from his convergents. A number of leading biologists have warmly welcomed the movement, finding in it relief from the strain of professing to be satisfied with Epiphenomenalism or Psycho-physical Parallelism and that subordination of biology to physics to which those doctrines owe their existence and which in turn they have emphasized.[1]

These biologists feel that the new movement sets free biology from the domination of physical science. Professor Jennings, for example, who hails with joy the doctrine of Emergent Evolution as the Declaration of Independence of biology, writes: 'No longer can the biologist be bullied into suppressing observed results because they are not discovered nor expected from work on the non-living parts of nature. No longer will he feel a sense of criminality in speaking of relations that are obvious in the living, for the reason that they are not seen in the non-living. Biology becomes a science in its own right.' And Professor Herrick rejoices that we need no longer maintain the pretence that conscious thinking makes no difference to our behaviour, that we could and would behave in just

[1] Among those biologists who have welcomed Emergent Evolution are Drs. C. J. Herrick, H. S. Jennings, G. H. Parker, W. M. Wheeler, R. S. Lillie and W. E. Ritter. In America the word 'organicists' seems to be used to denote those biologists who accept the principle. 'Emergent Vitalism' is an expression (proposed by Mr. C. D. Broad, I believe) to denote the biology of the kind that results from the acceptance of Emergent Evolution.

the same ways if consciousness were somehow excused from dancing an assiduous but futile attendance upon our brain-processes.

Other biologists, like Professor Julian Huxley, influenced no doubt by Bergson, but also by the 'emergence' movement, to which Bergson perhaps hardly belongs, no longer hesitate to speak of creative evolution.[1]

Many things, new relations, new qualities, new properties, including Life and Mind, are said to have emerged in the course of evolution; and other emergents of a more august nature are said to be promised or implied in the future.[2]

We must first make ourselves clear as to what is meant by Emergent Evolution; we must then inquire whether it is acceptable as one grand scheme of universal evolution, and, if not, what of it is acceptable. Finally, we must examine its bearings upon our central problem, the status of teleological causation.

It is not easy to give any simple statement of Emergent Evolution; for the leading exponents of it do not agree in respect of very fundamental points.[3] Many of those who express a general acceptance do not tell us just how much of it or in what sense they accept it. But I shall venture to

[1] Cp. the recent volume *Creation by Evolution*, edited by F. Mason (London and New York, 1928).

[2] E.g., the God or the Quality of Deity of Prof. S. Alexander.

[3] This is true especially of Professors Alexander and Lloyd Morgan, the two most thorough and careful exponents of the doctrine as one of universal evolution; the former in *Space, Time and Deity* (London, 1920), the latter in *Emergent Evolution* (1923), and *Life, Mind, and Spirit* (1926).

attempt a statement as concrete and simple as possible that shall fairly represent what seem to be the main features most widely accepted.

The story begins with a universe consisting of some physical forerunner of matter; we are not told anything definite of its nature. It is distributed in systems of very simple organization. The systems become more complex in various degrees; protons and electrons, let us say, become arranged in varied systems. These are the chemical elements, oxygen, hydrogen, sulphur, iron, etc. That is one great step. The qualities and properties of these so-called elements are emergents; they are new in the world on their first appearance; when they have appeared events different from all previous events become possible.

These elements combine with one another in various ways. Oxygen combines with hydrogen and the properties of a molecule of water emerge; water with all its peculiar qualities and properties is an emergent, a novelty. It consists only of oxygen and hydrogen; but, just because these are related to one another in a new way, the combination, H_2O, exhibits emergent qualities. The molecule of water is formed by the combination of hydrogen and oxygen alone; no other element or factor of any kind enters into it; but it is not hydrogen and oxygen; it is water. In the combination, potentialities previously latent in the hydrogen and oxygen have become actualized, with the consequence that the molecule of water exhibits qualities and properties not exhibited by, nor discoverable in, hydrogen and oxygen in their pure state. The new kind of relatedness is intrinsic to the system ; the emergence of the new qualities and properties does not depend upon new external relations of any kind.[1]

In similar way carbon combines with sulphur to form carbon bisulphide, chlorine with hydrogen to form hydrochloric acid, and so of all the multitude of chemical compounds ; the formation of each was an instance of emergence, was a creative synthesis of a system having qualities and properties that were new in the universe : consequently events of new kinds began to occur.[2]

After a period of such chemical emergence came a great step. Some system of atoms (chiefly O, H, C, and N, with perhaps P, S, and Cl,) attained a new degree of complexity, and Life emerged. ' Life ' is the word with which we sum up the peculiar qualities and properties that emerged on that occasion. There was nothing added from outside the system ; no new elements entered it, no new energies or forces played upon it from outside ; the new kinds of relatedness of the atoms and energies within the system sufficed. As these new intrinsic relations were established, the system began to exhibit the

[1] I follow Lloyd Morgan in his useful differentiation of the words ' quality ' and ' property ', to denote respectively the intrinsic features of anything and those which it manifests only under extrinsic influence ; and in general I shall make use of his carefully defined terminology.

[2] The new realist, who is also an emergentist (Prof. Alexander, for example), maintains that the peculiar odour which we experience in the presence of carbon bisulphide came into existence at the moment when the compound first was formed, though it had to wait many millions of years before there emerged a nose capable of appreciating that odour.

complex of qualities and properties that we call life—irritability, conductivity, power of growth, assimilation, metabolism, generation, regeneration, and all the rest.

After an interval, long or short, during which living systems became more complicated, the new complications at last rendered possible further new kind of relatedness and there emerged sentience, the forerunner of Mind. Further complications ensued (again with new kinds of relatedness), cognition emerged, and Mind for the first time appeared upon the scene. Then slowly those systems in which Mind had emerged increased in complexity of organization, and new forms of experience emerged. Thus the process continued up to the emergence of the higher forms of intellect and moral personality.[1]

In this story the feature of chief interest is the alleged successive emergence of Life and of Mind from a world of purely physical reality, a world of which large parts still remain perhaps in the primordial unevolved condition of nebular matter or whatever the original matrix may have been.[2]

Let us note at once that, if the scheme were acceptable in its main outlines, it would follow that teleological causation is but a disguised form of mechanistic causation, crypto-mechanistic, so to say, as is commonly assumed. For it would be absurd to attribute teleological or purposive intelligent action (which as we have seen implies always a considerable degree of organization) to electrons

[1] According to Alexander's version of the story this last step involves the emergence of the quality of Deity.

[2] Note 12. *Various Versions of Emergent Evolution.*

or ether or any other primordial unevolved stuff. And, as we have also seen, no conjunction of mechanistic events can yield a teleological event; truly teleological events cannot have emerged from any such conjunction.

I begin the criticism of Emergent Evolution by reciting what seem to be the essential facts in the history of the doctrine.

J. S. Mill had pointed out in the middle of the nineteenth century the logical difference between a mere aggregate or sum of entities and a synthetic whole. In the late years of the century the psychologist, Wilhelm Wundt, pointed to the fact that, though it was usual in psychology to speak of complex sensory qualities or complex presentations as formed by the fusion of simple sensations or sense-qualities, the complex so described was not merely an aggregate of such sensations; it was a whole that seemed to be more and other than the sum of the elements that were said to enter into its composition. Thus, when three notes are sounded together, we may hear what we call a clang or a chord; and, if a number of notes or chords be sounded successively, we may hear a melody. The melody is more than a mere sum or aggregate of notes; it is a whole of a richer, more significant kind that works as a whole upon the further course of our mental life. Or a series of words is uttered, and we hear not a mere succession of words but an intelligible sentence; the series of words has or evokes a meaning, which is other than, and more than, the sounds of the words, and more than a sum or aggregate of the meanings of all the several words. Our mental life affords a multitude of similar illustrations of a principle which Wundt proposed to describe as the principle of creative resultants. Other psychologists, recognizing the vast importance of this principle in our mental life, have preferred to call it the principle of creative synthesis.

Now J. S. Mill also had noticed some of these instances of mental synthesis and, observing that they are not to be described or explained in terms of the purely mechanical association-psychology which he had inherited from his father, he proposed to speak of them as illustrating a principle of mental chemistry. That proposal was the true origin of Emergent Evolution. At that time Atomic Materialism was still in vogue. But the physical scientists were beginning to realize that its simple mechanical principles were not adequate to the explanation or description of the facts of chemistry or of electro-magnetism. This truth, which perhaps should have been obvious from the first, has been more and more clearly recognized. The molecule of water has properties that cannot be regarded as a mere sum or resultant of the properties of the atoms of hydrogen and of oxygen that are contained in it. It is a synthetic whole the properties of which depend upon the presence of its atoms of H and O, but those atoms are in reciprocal causal relations ; the whole is a system the properties of which depend, not only upon the presence of the atoms, but also upon the peculiar reciprocal active or causal relations between them. Something similar seems to be true of all chemical compounds and of many physical wholes, such as an electro-magnet field, or a soap-bubble.

We have then to recognize true synthesis in both the mental and the physical realms ; in both realms synthetic complexes are more than aggregates of the parts which in some sense may be distinguished within them.

J. S. Mills's use of the expression 'mental chemistry' had suggested some analogy between such mental and such physical syntheses. G. H. Lewes had already used the word 'emergence' to describe the production of physical or chemical synthetic properties ; and there was a third realm of uncertain status, the realm of vital phenomena, the phenomena of growth and morphogenesis of living things. In Germany a few years ago the psychologists of the *Gestalt* school proposed the word *Gestalt* (best translated 'configuration') to denote the peculiarities of synthetic wholes in both the physical and the mental realms. And in England about the same time the word 'emergence' began to be used as a word descriptive of the process of synthesis, and 'emergent' as a word denoting the properties or qualities peculiar to any synthetic whole. The German and the English authors alike imply that the principles of synthesis are similar for the two realms, and further imply that all the peculiar phenomena presented by living organisms are to be regarded as instances of the principle of configuration (*Gestalt*) or of emergence.[1]

[1] Thus Dr. W. M. Wheeler writes : 'Now emergence in the following pages signifies neither the manifestation or unveiling of something hidden and already existing, as in the common and the entomological denotation of the word, nor some miraculous change—but a *novelty of behaviour* arising from the specific interaction or organization of a number of elements, whether

A further step was to synthesize emergence with evolution (an instance of creative synthesis in the mental realm) and the doctrine of Emergent Evolution emerged.

The question before us is, then, whether the identification of synthesis in the physical and mental realms implied by the words 'configuration' and 'emergence' is valid. I suggest that it is not valid; that the words are used to denote two types of synthesis that are fundamentally different and distinct; and that by the use of the words 'configuration' (*Gestalt*) and emergence' it is falsely made to seem that creative synthesis (which undeniably occurs in the mental sphere) occurs also in the physical.

Let us notice the chief peculiarities ascribed to emergence as a natural process. And, in accordance with the sound principle that we seek the distinctive marks of any type of thing or event in those instances that display them most clearly, we may consider instances of emergence or creative synthesis in the mental sphere. It is claimed that each instance of emergence is creative of real novelty, of some new

inorganic, organic, or mental, which thereby constitute a whole, as distinguished from their mere sum or resultant.' (*Emergent Evolution and the Development of Societies*, New York, 1928.) And Mr. Broad has proposed the term 'emergent vitalism' for the view of living organisms to which Emergent Evolution naturally leads; namely, a view which is vitalistic in that it denies the adequacy of the laws of physics and chemistry to interpret the events that are characteristics of living organisms, but differs from all forms of vitalism that postulate in organisms any peculiar qualities, relations, or modes of happening that are not emergents from physical events.

quality or property of a type that did not exist before the emergence; that it is non-mechanistic and unpredictable before the event on the basis of any knowledge, no matter how complete, of the things or events that enter into the synthetic event; and that the new kind of relatedness, or the new quality or property, has causal efficacy, makes a difference to the further course of events.

It is claimed that the recognition of the principle of emergence renders Emergent Evolution vastly superior to such a universal scheme of evolution as Herbert Spencer's. The latter, it is said, dealt merely with aggregates and sums and resultants; every event was regarded as in principle predictable; it explained the higher by the lower; for it, Life was nothing more than a more complex conjunction of qualities and events common to the inorganic realm; and Mind or consciousness was a mere by-product, something without causal efficacy, a kind of phosphorence that appeared when the physico-chemical processes of the brain attained a certain complexity, without making any difference to the go of those events. And the chief superiority claimed for the new doctrine is that it recognizes the reality and causal efficacy of new qualities and kinds of relatedness, especially of that relation we call cognitive, within the organic realm. As Lloyd Morgan puts it: 'Emergent Evolution urges that the "more" of any given stage, even the highest, involves the "less" of the stages which were precedent to it and continue to co-exist with it. It does not interpret the higher in terms of the lower only; for that would imply denial of the emergence of those new modes of natural relatedness which characterize the higher and make it what it is.' That is to say, when Life and Mind have emerged, they involve new qualities and new forms of relatedness which thereafter do make a difference to the go of the events from which they are emergent.

Now all this is manifestly true of the mental realm. Consider the production of any original work of art, say a melody, the Marseillaise. Rouget de Lisle, stirred by an aspiration which he shared with many of his fellows, felt the need of giving it expression in music. He groped more or less blindly towards the goal that he vaguely conceived, no doubt with something of what is called 'trial and error'; and then the Marseillaise emerged. It was strictly a novelty, something new in the history of the universe; it comprised new kinds of intrinsic relatedness and as a whole possessed a quality that was new and unique. Its emergence was non-mechanistic in the true sense; not only could it not have been predicted, but also it could not have emerged without the forward-looking desire to make actual that which was vaguely conceived as a possibility of the future; that is to say the emergence was a teleological event. Further, when this novelty had emerged, it was effectively related to the events from which it had emerged; it had causal efficacy and contributed mightily to the production of consequences of the utmost importance.

All this is true in lesser degree of the emergence of any new melody, no matter how simple; and still more obviously true of the production of great works of art, an oratorio by Handel, a sonata or a symphony by Beethoven or Brahms, a landscape by Turner, an ode by Wordsworth or Keats. It is true also of every instance of original production in literature and science ; it is true in some degree of every genuine instance of the solving of a problem, every original discovery or invention, however humble or exalted. Thus the artist of the kitchen produces a dish that procures us a new experience in the sphere of flavours. A thinker, like Lloyd Morgan, launches a new way of conceiving some thing or event, say the way of Emergent Evolution ; and there is something strictly new in the world, a true emergent, one that may have further consequences, such as the delivery of this lecture, or perhaps a new trial at Dayton.[1] A mathematician achieves the solution of a hitherto unsolved problem, say, or, after long labour directed to the achievement of the goal, an Adams calculates the position of an unseen planet and, turning his telescope upon the place calculated, sees 'a new planet swim into his ken '.

All these seem to be real instances of creative synthesis. But perhaps we should make a distinction between instances of the type of the synthesis of sense-qualities, and those of the kind illustrated by our other illustrations. It is instances of the latter type that are the most indisputable instances of creative synthesis, of emergence of novelties that could not have been predicted ; and these emergents are the products of teleological causation, of intelligent activity striving towards a goal. The events

[1] It is worth noting that in this, as in most instances of intellectual creation, the emergence is a process that involves the co-operation of many minds.

are non-mechanistic, not merely because they could not have been predicted, but because they can be neither explained nor described without taking into account the mental reference to the future involved in them, the foresight at first vague but gradually defining itself, the striving to make real that which is foreseen and to foresee more fully, more adequately to the governing purpose of discovery or creation.

Let us now examine alleged instances of emergence in the physical realm, instances of physical or chemical synthesis. 'The classical example,' says Professor Wheeler, 'is, of course such a chemical compound as H_2O, in which hydrogen and oxygen combine under certain conditions and in certain proportions to form a liquid emergent, water, exhibiting a very different behaviour (properties) from that of either of its gaseous components.'[1] It seems here to be implied that the emergent property is liquidity. But that is not to be taken seriously; under certain physical conditions hydrogen and oxygen are themselves liquids and H_2O a gas. Still, H_2O is a product of synthesis, and it manifests certain properties not manifested by any other chemical substance, properties which enable it to play a unique role in nature; e.g., its strange property of expanding instead of contracting in volume as it approaches freezing-point, its properties as a solvent and its property of condensing to a liquid and solidifying to crystals under certain conditions of pressure and temperature. All these and many other properties are peculiar to water and make it

[1] Op. cit.

what it is. But are they emergents in the same sense as the mental syntheses we have cited ? Are all or any of them strictly unpredictable ? Do they imply ultimates or indefinables and principles or natural laws peculiar to themselves ?

It is the aim of the physical sciences to express or interpret all physical events in terms of the smallest possible number of postulated indefinables. Emergent Evolution asserts that every property peculiar to any chemical compound is such an indefinable. I believe that few if any physical chemists would agree to this ; they would maintain that, though the indefinables of Atomic Materialism have proved inadequate, and though physical scientists are not yet agreed upon the indefinables that are indispensable, yet we may reasonably assume that the postulation of a small number of indefinables and a few highly general laws will eventually render it possible to account for all the properties of inorganic substances, to explain mechanistically all physical events, and in principle to predict them. This is the programme which physical scientists hold before their minds ; and their many successes in the working out of this programme seem to justify the belief that in principle it is one that may be achieved. Emergent Evolution makes the unwarranted assumption that any such programme is impossible of achievement, that each chemical compound expresses in its peculiar properties empirical laws that are ultimate, that cannot in principle be derived from some more general laws.[1]

[1] Mr. A. D. Broad (in *Mind and its Place in Nature*) maintains this emergent doctrine very confidently. He writes : 'The law

There is a very real and important difference between a chemical and a mental synthesis, a difference which in itself suffices to render improper the subsumption of them under the single principle of emergence. We see this clearly if we compare typical instances of the two principles. A molecule of water illustrates the physical principle of *Gestalt* ; a musical chord as heard the pure mental principle of creative synthesis. In each case the whole is more than the sum of its parts ; in each case the parts are determined by the whole and the whole by the parts in reciprocal effective relatedness. But there are great and fundamental differences. The

connecting the properties of silver chloride with those of silver and of chlorine and with the structure of the compound is, so far as we know, a unique and ultimate law. . . . It is a law which could have been discovered only by studying samples of silver chloride itself, and which can be extended inductively only to other samples of the same substance.' Now this, of course, is a question for the physical chemists to answer. Unfortunately none of them seem to have pronounced hitherto upon the principle of emergence in their realm. But it is clear that, in the instance cited by Broad and many other similar ones, many of the properties of the compound could have been predicted from knowledge of silver and chlorine and of such compounds as silver bromide, etc. If no samples of silver chloride had ever been available for study, its weight, its crystalline form, its solubility, its general properties as a metallic salt, could have been predicted. It is for the emergentists to point to specific instances of what they regard as emergent qualities in the physical realm ; and this they neglect to do. It is implied in their doctrine that in such elements as chlorine and silver there are properties that remain latent and undiscoverable until the two elements enter into combination ; and that then these latent properties are manifested in the emergent. In view of the fact that chemists have been able to predict the existence and many of the qualities and properties of elements never actually observed, this claim of the emergentists seems very ill-founded.

atoms that enter into the composition of the molecule existed independently of one another prior to that entrance; and they can be recovered from it, separated from one another, apparently unchanged. Though we loosely speak of the chord or complex clang as formed by the fusion of sensations, this way of speaking is strictly incorrect. The sense-qualities which by introspective analysis we may recognize in the clang or chord did not exist separately before the experience of the clang or chord; nor can they be recovered from it and set apart as sensations, existing independently of one another. Again, the water is a combination of elements in a synthetic whole, the molecule, which having been formed, may continue to exist indefinitely without further change: the hearing of the clang, on the other hand, is a synthetic activity and the clang exists or subsists in any sense only so long as that activity continues.

I submit, then, that, though there are points of analogy between chemical and mental synthesis, we have to deal with two radically distinct principles which cannot properly be subsumed under one principle of configuration (*Gestalt*), or of emergence. I submit that in the physical realm there is no true emergence of novelties; that consequently there has been no emergent evolution of or in the physical realm.[1]

[1] Those exponents of Emergent Evolution who (like Prof. Alexander) are also neo-realists claim to find true instances of emergence in the physical realm; for they regard the various sense-qualities (colour, tone, warmth, coolness, odour, tickle, smarting, and so forth) as properties of physical objects. If this peculiar neo-realistic doctrine were tenable, these would be

I submit that, though it is true that the principles of Atomic Materialism or strict mechanism are inadequate to the interpretation of chemical and many other events of the physical realm (because true synthesis and configuration occur in it), the recognition of this fact does not justify the assumption of emergent evolution in the physical realm.

Greatly daring, I will go further and submit, not only that no emergent evolution occurs in the physical realm, but also that in that realm there has been no evolution in any proper or significant sense of the word. The term evolution was applied to the history of the realm of living things and has been extended to the inorganic sphere under the influence of a supposed analogy. I submit that there is no true analogy between the histories of those two realms and that the use of the one word ' evolution ' to denote these two histories is perilously misleading and should no longer be countenanced.

This a grave heresy for any man of science to harbour ; yet I feel sure that it is one of those heresies that are destined to become orthodox doctrine.

true emergents and Emergent Evolution would to this extent be true of the physical realm. But it is not tenable. It flies in the face of a multitude of facts, as its many critics (e.g., C. A. Strong, Durant Drake, and Lloyd Morgan) have shown. And the whole doctrine has been devised just in order to bridge, or to diminish, the gap between the physical and the mental by importing something of mind into the physical realm. The neo-realists desire to avoid, by means of such importations, what Whitehead calls ' the bifurcation of nature '. But, if nature is really bifurcated, we gain nothing by disguising the fact with elaborate fictions.

The so-called evolution of the physical universe is a highly speculative story. We are told that we must postulate some primordial condition of matter and energy preceding the emergence of matter, perhaps preceding the differentiation of matter and energy, or of matter and ether. There is no agreement as to what this primordial condition may have been ; and it is impossible to conceive the physical universe as having suddenly begun to be or as having existed without change in its primordial undifferentiated state and then begun to evolve. That is to say there is no reason to believe that at any remote time changes of the kinds now going on did not go on.

The assumption which most gives plausibility to the story is that the material universe is an island in an infinite ocean of ether, and that the energy of which it was at one time possessed has been in process of dissipation by radiation into infinite space, according to the laws of entropy. Suppose this to be true and that our island of matter, the stellar universe, is approaching a condition in which all the matter will be collected in one solid mass at zero temperature. This history, when completed, would be one of a process of change in a certain direction. But would it be the history of an evolution ? In what sense would it be analogous to organic evolution ? I see no analogy.

The one is a process of running down, of loss of energy and potentialities. The other is a process of running up, of progressive realization of potentialities. The characteristic feature of the one is degradation and dissipation of energy ; of the other, concentration of energy and the raising of it to higher potentials.[1]

Organic evolution has been, according to all accounts, a gradual building up of organizations that are more and more complex and efficient ; a differentiation and specialization of functions that contribute more and more effectively to the attainment of one end, namely, the maintenance and propagation of life under whatever physical conditions may obtain. There is nothing analogous to this in the history of the physical realm. There are, as we have seen, no functions in the physical realm ; the word is not applicable there, unless we postulate a Designer and Creator with some end in view. And with due deference to those philosophers who like Whitehead tell us that the whole of nature consists of organisms, I venture to question whether the word ' organism '

[1] It is not clear to me whether the process of degradation and dissipation of energy must result in the aggregation of all matter in one mass. But that is a question of very minor importance. The essential question is whether the process of degradation and dissipation can properly be called evolution. It would seem less inappropriate to call it a devolution. The same considerations apply whether we consider the physical universe as a whole or parts of it. Can the cooling and other processes by which the moon has reached its present condition be called an evolution ? Is the burning to ashes of the logs on my hearth, or of my house, an evolution ? If we seem to find a certain propriety in speaking of the evolution of the earth prior to the appearance of living things upon it, it is because we implicitly regard the process of condensation, the formation of the geological strata, of ocean and air and land and river and soil, as the preparation of it to be the abode of living things and of mankind. An alternative theory of the course of change undergone by the physical universe is that matter has been undergoing destruction and dissipation into infinite space by radiation in the form of energy. It would be still less proper to call this course a process of evolution. Cp. Note 9, on the *Theory of the Dissipation of all Matter*.

can properly be applied to inorganic things that are not the products of design or of telelogical causation. In other words, in the physical realm, the machines constructed in the service of some purpose are the only physical structures that can properly be said to have functions and to be, in some sense, organisms. And even such extension of the word is of questionable validity. The word efficiency is properly applied to machines and their parts. But it cannot be applied to any other physical thing or process or event

The only sense, then, in which the inorganic realm may be said to have evolved is in the sense of having become more complex. But, accepting still the history of it as a course of change resulting in the consolidation of all matter in one lump at zero temperature, would that be a history of increasing complexity? I do not know; and the answer to this question is not of prime importance, for even if it were a history of increasing complexity, such increase alone would not make it an evolution.

True evolution, organic evolution, is a progress of organization in respect of complexity and efficiency; and complexity is subordinate to efficiency. Mere increase of complexity of organization that brought no increase of efficiency would not be evolution; any instance of that sort in the organic realm would be, by common consent, regarded as one of degeneration rather than of evolution. Perhaps there is no such instance in nature; though something of the sort is perhaps realized in monsters. Let us take then an imaginary case. Suppose one of the hideous idols with ten arms conceived by certain artists were realized in a living form. That would be increase of complexity, but not of efficiency ; and it would be an instance of degeneration rather than of evolution.[1]

But we have made a concession to the physical evolutionist in supposing that our island of matter is approaching a condition of consolidation. In the light of the Millikan ray and of many other considerations any such fate seems very problematical. It seems possible that the matter of the universe is destined to surge to and fro indefinitely, becoming wrought up into more complex systems with varying conditions as regards the quantities of active and potential energy ; perhaps with dissolution of matter into energy occurring as often and as abundantly as changes of the condensation type.[2]

Let us look at the alleged physical evolution more closely. Two alleged instances of it seem to have better claims than any others for critical examina-

[1] Lloyd Morgan abstains, wisely perhaps, from giving us any definition of what he means by evolution. Yet his words imply that he means increasing efficiency ; as when he writes : ' Progressive advance in evolution is here my theme.' But mere increase of complexity is not ' progressive advance '.

[2] It has recently been suggested by a leading physicist, Sir J. H. Jeans, that Laplace's great speculation is possibly true : that there was at one time a cloud of diffused matter ; ' Out of the cloud nebulae condensed, out of the nebulae stars, out of the stars planets, and out of the planets satellites. . . . Where did the primeval cloud come from ? Possibly from the fifth dimension. Jeans considers that the difficulty of explaining the shape of the spiral arms in the great nebulae may be solved by the discovery that the centres of such nebulae are taps through which matter pours from some other universe into ours.' (I cite from the *Manchester Guardian* of August 3rd, 1928.) Would this be physical evolution ? It would seem to be a case of immergence of matter rather than emergence. Cp. Note 9.

tion ; the formation of planetary systems and that of complex molecules. Suppose it to be true that our planetary system was formed by condensation from a nebula, spiral or other. In what sense is that an evolution ? In what sense is the present state superior to the nebular or any prior state ? In what sense has there been ' progressive advance ' ? Is it more efficient or more complex ? It is more efficient only in the sense of being better suited to serve as the abode of living things. Apart from life, it is more complex only in so far as it contains more complex forms of matter, or forms of matter in more complex relations with one another.

It would seem then that the history of the physical universe has been in any sense an evolution only if more and more complex forms of matter have been produced in it. Is there then any good ground for believing that, apart from living things and their chemical products, there has been such increase of complexity of chemical compounds ? Is there reason to believe that any inorganic molecules now existing could not have existed and did not exist at times indefinitely remote ? I do not know of any such reason. It is true that, if our solar system or any other one was at one time at a very high temperature in all its parts, such temperature may have been inconsistent with the existence of the more complex chemical compounds and that, as the system has cooled, there has been some general condensation of matter in more complex molecules. But, as was said above, a mere cooling and consolidation, does not constitute an evolution in any useful sense of the word ; to call it such is merely to produce confusion through loose talk. I submit that there is, apart from living things, no chemical compound that might not have been formed at any time in the history of the physical universe, if certain physical conditions of temperature, pressure and so forth had existed. The same cannot be said of any organism known to us.

Here we come upon the essential difference between organic evolution and merely physical and chemical changes. If, in the course of any of the changes that have taken place in the physical realm, the requisite conjunction of forms of matter and of physical conditions (such as temperature, pressure, moisture, etc.) had occurred or been in any way produced, any chemical substance (with the exception, if any, of those dependent for their formation upon living organisms) might and would have come into existence forthwith. No course of progressive development was a necessary prerequisite of the formation or, if you will, the emergence of such compounds. The same cannot be said of any living organism, whether a man, a frog, a *vorticella*, or even an *amoeba*. Each kind of organism is the product of a long process of cumulative change, no step of which could have been other than it was. It is true that we can point to instances of similar organs that seem to have been evolved independently along different lines; but that is not true of whole organisms. Each type of organism is a unit that has its own evolutionary history; and only that course of change could have resulted in that particular organization. Each organic type, in short, bears the marks of its history; as Bergson puts it, it carries its past along with it; it has a true history, the history of its evolution. It is true that an intelligent guiding agent might have abbreviated the period of time occupied by the evolution, in some such ways as the breeder uses in producing by selection a new type; but such guidance could not have dispensed with any of the stages of the process. They were necessarily successive in the order of their actual occurrence.

Now it may be said that something analogous to this is true of the formation of chemical compounds; that the formation of the more complex presupposes the existence of the less complex. It is true that a chemist, desiring to produce synthetically a complex compound, generally proceeds by bringing together more simple substances. He takes, let us say, two salts and brings them together at a certain temperature; there results a compound more complex than either one. Well, my point is that the history of the salts thus brought together is perfectly indifferent to the success of the procedure. The molecules of one (or both) of them may have existed in the rocks or in the sea for a hundred million years, or they may have been made in a test tube on the morning of the experiment. They may have been obtained by a process of synthesis or by one of decomposition of a more complex compound. The difference of history makes no difference to the present process.

Here is the root of the difference between organic evolution and physico-chemical change in the direction of greater complexity of structure. The former implies memory; the other does not. There can be no true evolution without memory. The word ' memory ' is here used in the widest legitimate sense ; it implies stability of organization in spite of change, a stability which enables it to survive change and to incorporate effects of change within itself. It is a stability that involves both elasticity and plasticity ; both in a degree unknown in the physical realm and never combined in any physical structure. In the physical realm elasticity and plasticity are opposite and incompatible properties ; in proportion as a physical structure is elastic, it lacks plasticity, and conversely.

We see this unique stability of vital organizations, this combination of elasticity with plasticity, illustrated in a hundred ways. We know that certain species have persisted unchanged through many millions of years. It is probable that certain of the Protozoa (perhaps *Amoeba proteus*, that viscid speck of protoplasm undergoing perpetual changes of form and substance) have remained essentially stable or unchanged through immense periods of time ; that is to say, in spite of perpetual changes going on in all individuals, the organization of the species has persisted unchanged. We see another but less simple illustration in each instance of recapitulation of phylogeny by ontogeny. The growing organism repeats the history of the evolution of the species. And, even if the course of recapitulation undergoes distortion (as by the development of organs of special embryonic function, such as the placenta, or by gross mechanical interference from outside) the stability of the organization of the species, of its memory, asserts itself, and the normal form is evolved. In the inorganic, there is nothing comparable to this stability in spite of change that we denote by the word ' memory '. Shift a planet from its course by a hair's breadth, and you change all the further history of that planetary system. Change the relations, spatial or other, between the constituents of a molecule, and it is no longer the same kind of molecule, but another.

The stability of vital organization in spite of change, a stability which enables changes to be incorporated in the whole without loss or radical alteration of the pre-existing organization, this stability peculiar to living organisms is an essential condition of evolution. It may be impossible to understand or explain it, but it is a fact which we must accept with that natural piety so often demanded of us by the emergent evolutionists.

We are often told that, of all things of the physical realm, crystals are nearest to organisms in respect of their formation and properties. Here, then, if anywhere, we might expect something analogous to true evolution. But the same kind of crystal may take shape in a multitude of different ways. Of a hundred similar crystals each may have its own peculiar history, be the product of a process very different from that of which any other is the product. There is no repetition of phylogeny by ontogeny ; indeed there is no phylogeny in the matter ; there is no memory and therefore no evolution.

A demon chemist might have produced any kind of crystal at the dawn of the sidereal system (if there was such a dawn) given a sufficient power of controlling physical conditions. Suppose also that his knowledge and powers of manipulation of material things had been so great that he could have arranged atoms, molecules, and electrons in a system exactly resembling the material structure of a fertilized human ovum. Would that system, given suitable material environment, have developed into a man? The emergent evolutionist says: Yes. I say: No; there is no good reason to suppose that it would have done so. For the system would lack just that essential part of the organization which is memory. Man or any other organism as it now exists could have been produced only by the series of steps which have actually occurred in the evolutionary process. We are able to infer the past course of the evolution of any organism from observation of the manifestations of what the organism is, especially the manifestations afforded by ontogeny. This is not true of the crystal or of any physical thing known to us; every merely physical thing is indifferent to its past history; its past cannot be inferred from the most detailed knowledge of its present structure, qualities and properties. Take two similar atoms or two similar planetary systems. One may have been formed in the course of disintegration of a more complex system; the other in a process of condensation and complication of simpler systems. Yet now, in spite of the wide differences of their modes of formation, they are alike in every way.

The truth seems to be that biologists, concentrating their attention on the physico-chemical or material structure of organisms to the neglect of their mental life, have commonly assumed that their peculiarities are wholly the expression of great complexity of chemical constitution, and that organic evolution has consisted in increase of such complexity. If it were possible to regard the whole of a complex organism as a single molecule, the common assumption would have a certain plausibility. But this view, though it has been suggested, is quite untenable.

Instances of true evolution are afforded by inorganic structures such as ships, steam engines, internal combustion engines, and machines of many kinds. And such instances may seem to loose thinkers to afford support to the view that evolution occurs in the inorganic realm. But they give no such support. All complex machines, like all political and legal and religious systems, and like all human institutions, have been produced by emergent evolution ; and there is good ground for believing that in all cases of real evolution, Mind, that is to say, intelligence, purpose and memory, have played the essential role. It is, I submit, impossible to point to any clear instance of evolutionary progress of which this can be denied.

CHAPTER VI

THE ALLEGED EMERGENCE OF MIND

IT is, we have seen, a leading feature of Emergent Evolution that it claims to describe the evolution of Mind from the physical realm, and to render this account acceptable by recognizing the principle of emergence as valid in both the physical and the organic realms. Let us now grant, for the purposes of the argument, that assumption which we have found good reason to deny, namely, the assumption of the reality of Emergent Evolution in the physical realm, and let us examine the alleged course of emergence of Mind.

We encounter here a peculiar difficulty. Hitherto only four schemes of the emergence of mind have been elaborated in detail: that of Alexander, that of Lloyd Morgan, that of E. Noble, and that of Drs. Strong and Drake. These four schemes differ in respect of certain fundamental assumptions; and a thorough examination would have to deal with each of them separately. This would be a very tedious procedure. I adopt the plan of choosing one of them for examination and appending notes upon the other schemes. There can be no doubt that, for the majority of biologists who accept Emergent Evolution, Lloyd Morgan's account, purified by the extrusion of certain inconsistencies, is the most acceptable, as it is the most thoroughly elaborated; it may be regarded as the most authoritative. Dr. Lloyd Morgan is the most distinguished of comparative psychologists, and has a wide familiarity both with many branches of science and with philosophical modes of thought. He, if anyone, should be able to render a plausible account of the emergence of Mind. Further, his account, is very similar to the account rendered by Strong and Drake, who, like Morgan, are critical realists.

The first step in the emergence of Mind was, we are told, the emergence of sentience, a passive suffering of sense-impressions, anoetic sentience that has no causal efficacy, no reaction upon the chemical events of the brain from which it emerges.

Let us note without delay that this contradicts the fundamental principle of emergence, the principle of effective relatedness; it involves a falling back on Epiphenomenalism. Yet the overcoming of Epiphenomenalism, with all its imperfections, not to say absurdities, the banishment of it to the limbo of historical curiosities of speculation, is one of the chief merits claimed for the emergent principle.

Further, anoetic sentience is only what is called a limiting conception, arrived at by stripping away in imagination all the characteristic modes of mental functioning other than this passive quality of experience reduced to its lowest terms. If we admitted the validity of conceiving anoetic sentience to occur, we should soften the transition from matter to Mind; for such sentience would be neither one nor the other, but something in between, something having the quality of Mind without its characteristic modes of causal efficacy. Yet the postulation of its emergence from physico-chemical synthesis is a much more questionable assumption than the postulation of the appearance of material systems having properties previously not realized, the appearance, for example, of solidity in a universe previously containing only fluid matter. We cannot positively assert that sentience did not or could not have emerged; but we can assert that, if it so emerged, its emergence remains perfectly unintelligible and inexplicable after the event; whereas the emergence of solidity, even though it were unpredictable before the event, becomes intelligible and explicable thereafter.

In fact, this alleged emergence of sentience, the supposed first step in the evolution of Mind, is so hard to swallow that all schools of emergence, both the neo-realists who follow Alexander and the critical realists who follow Lloyd Morgan balk at it and devise ways round, ways that shall avoid the appearance of swallowing so tough a morsel.[1]

However, let us grant the emergence of sentience; and we are still far from the emergence of Mind. It is generally agreed that all mental activity has three distinguishable aspects, knowing or cognition, feeling, and willing or conation. If the emergence of Mind is to be acceptable, we require some plausible account of the emergence of all three aspects.

Cognition implies reference to an object. Of such reference, alternatively called by Lloyd Morgan 'projicience', he gives different accounts in his two volumes of Gifford Lectures. In the earlier,[2] reference emerges at a late stage of animal evolution,

[1] Cp. Note 12, *Various Versions of Emergent Evolution.*
[2] *Emergent Evolution* (1923).

at the stage when the sense-organs for reception of impressions from a distance, the distance-receptors, have been organized. And its emergence occurs when sentience becomes complicated by simultaneous revival of sense-qualities formerly experienced. Up to this late stage, we are told, passive sentience without effective relatedness or causal efficacy, together with some vague enjoyment, equally without causal efficacy, was all of Mind that had emerged. But, with the complication of sentience by revivals of formerly experienced sense-qualities, reference or cognition emerged; and then conscious guidance of behaviour began. We are told also that this story of emergence of reference is repeated in the life-history of each animal that attains the level of reference and conscious guidance. Thus the young chick, making its first peck, acts in a mindless fashion; the behaviouristic account of the process would be adequate. But, when it has pecked at good- and ill-tasting substances, its visual sentience becomes complicated by revivals of olfactory sentience; it then acts with 'reference' and conscious guidance.[1]

[1] 'Is there a stage in the individual development of an organism in which consciousness is eventually emergent when there are sensory presentations that as yet carry no meaning? From the point of view of emergent evolution, there is such a stage—one at which a behaviouristic interpretation of that which happens is adequate and sufficient.' And such anoetic sentience without reference or meaning is attributed to the young chick pecking for the first time at a grain of corn. 'Has the presentation, as something for the first time given, initial reference to something else or something beyond? My own reply is that, in such a case, there is no such initial reference, that conscious reference only derivatively begins when there is revival of such experience as the little bird has already and individually gained

In the later volume [1] Lloyd Morgan, perhaps owing to criticism of this account, modifies it radically. In this later account reference does not emerge ; it is given in the very constitution of things as the natural correlate of causal influence.[2] In this second account the place of reference is taken by 'prospective reference'—this special kind of reference is now the emergent that introduces conscious guidance of behaviour. So far so good. Let us be grateful for this recognition of foresight of the future as essential in all mental guidance of action. But what makes the reference prospective ? No light whatever is thrown upon this fundamental question. Prospective reference implies appreciation of temporal relations, as proficient reference implies appreciation of spatial relations. And neither of Lloyd Morgan's two accounts throws the least light upon the emergence of these functions, fundamental as we have seen in all intelligent action.

in the course of pecking and other modes of behaviour on prior occasions. . . . All such reference, when it comes, is derivative from previous experiences in the individual life.'

[1] *Life, Mind, and Spirit* (London, 1926).

[2] This is to be understood only in the light of Lloyd Morgan's acceptance of the Spinozistic doctrine that the physical and the psychical are two attributes or aspects of all reality. According to this later account, 'reference' is present throughout the physical realm and in the simplest animals. Thus : 'Distinct from either concomitance or emergence, is the concept of reference as that which in mental regard is the reciprocal counterpart of influence in physical regard. Why this should be so—nay more, why there should be either influence or reference in the realm of nature—I know not. Both are given.' Further, we are told that 'even so simple an animal as amoeba has something, however rudimentary, of the nature of enjoyment, and something, however incipient, of the nature of reference to its environment. Thus far we do suppose that where there is life there is also mind, though it may be a very simple form of mind.'

The plain fact is that we cannot plausibly generate reference from anoetic sentience, nor prospective reference from reference pure and simple, by adding together bare sense-qualities, not even if we use the magic word 'emergence'. Dr. G. F. Stout has sufficiently exposed the inadequacy of all such attempts; and I need not repeat his criticism in detail.

This doctrine of the genesis of cognition by the conjunction of sense-qualities is the old discredited associationist doctrine of perception, modified by insistence on the principle of emergence or creative synthesis. It has long been current in the special form of the doctrine that all our knowledge or awareness of spatial relations arises from the complication of sense-qualities of the several special senses by sensations excited through the 'muscular' or kinesthetic sense; the doctrine that our spatial awareness, highly differentiated in three dimensions, is given or emerges or is creatively synthesized when visual (or other special) sense-qualities are complicated by the extremely obscure and undifferentiated qualities of the kinesthetic sense.[1] This doctrine has never been plausible; and it does not seem to be appreciably strengthened by the alleged extension of the principle of emergence to the physical realm; and such extension is the essence of Lloyd Morgan's attempt to render this doctrine acceptable.

[1] I have criticized it in my *Outline of Psychology*. It is essentially the doctrine of Strong and Drake. They make much of the fact that in visual perception the sense-organs are focused upon the object seen. But no amount of bodily movement directed upon the object can, in itself, convert a mere conjunction of two forms of passive sentience into a cognition, a conscious reference to an object in space.

I submit that the emergence of the cognitive function has not been made acceptable ; that rather we must agree with Dr. G. F. Stout in maintaining that 'the primitive mind directly apprehends a primary sensible, and in doing so refers it to a source', i.e. to some object however vaguely conceived ; that there is at the lowest level of mental development 'immediate knowledge of the sensible as incomplete' ; that 'the mind starts with some general apprehension of the unity of the world sufficient to enable it, when occasion arises, to expect and seek connections not yet disclosed. . . . If we are not quite gratuitously to place an impassable gap between the earlier and the later stages of mental development, one must assume that it is present in however indeterminate a way, from the beginning' [1] ; that, in short, reference or cognition cannot be legitimately regarded as evolved or as emergent out of some events or some functions that have nothing of the nature of cognition ; that rather it must be accepted 'with natural piety'.[2]

Let us remind ourselves that intelligent perception involves appreciation, not only of spatial relations, but also of temporal and of causal relations ; all of which Lloyd Morgan's account neglects entirely. Let us notice that Lloyd Morgan rightly asserts 'Just how guidance arises, what is its psychological accompaniment, is perhaps the crucial question in

[1] I cite from an article in *Proceedings of the Aristotelian Society* (London, 1913).

[2] This view, it will be noticed, accords with Whitehead's attribution of vague apprehension of causal efficacy to primitive minds.

the whole range of evolutionary advancement.'[1] That he then gives us two very different accounts of this crucial instance of emergence, neither of which is in the least adequate or even plausible. In both accounts the complication of sentience by alleged revived sense-qualities is the essence of the process.

In the one account such complication leads to the emergence of reference and of conscious guidance at an advanced stage of animal evolution. In the other it leads to prospective reference; simple projicient reference being represented as a fundamental aspect of all reality, which, therefore, together with enjoyment must be attributed to the simplest forms of life.

Lloyd Morgan's account of the emergence of Mind deals only with the emergence of reference or cognition. It is true that this is a crucial question. But, if the emergent account of this step were satisfactory, we should still have the equally difficult problem of emergence of conation, of impulse or urge towards a goal. We have already seen that the various attempts to equate such conative impulse or urge with such physical processes as the return of a disturbed system to equilibrium are wholly lacking in plausibility, that they ignore the essential

[1] The crucial importance of the emergence of cognition Lloyd Morgan clearly recognizes in other passages also. Facing it, we are, as he says, ' at a crucial parting of the ways. . . . One route leads to the view that mind is emergent in the course of evolutionary history, the other path leads to the view that mind is not emergent ', that is to say, to the view here taken that mind is utterly disparate from matter, that its organization and functions are wholly unlike physical structures and cannot be regarded as having been evolved from the latter.

peculiarities of conative events. Given spatial, temporal and causal reference, intelligent conative events imply in addition this hormic factor.[1] Neither Lloyd Morgan nor any other emergentist has attempted this problem; therefore we cannot pretend to criticize the emergent treatment of it. Yet such treatment is required. Lloyd Morgan having recognized 'planful activity' and the peculiar experience, the inner urge, that is of the essence of conation,[2] would, I presume, not be content to equate conation with any purely physical process; he would regard it as emergent at some level of animal evolution, a level presumably posterior to or higher than that of the emergence of reference; for reference is implied or presupposed by conation. For, as I insisted in an earlier lecture, although various authors use the expression 'blind impulse' or 'blind conation', we have no warrant for such usage, unless we merely mean relatively blind.

In our own experience such relatively blind conation is not uncommon. It not infrequently happens that I leave the chair where I am writing to perform two errands, to fetch, perhaps, a book and a pamphlet, both of which, as I foresee, I shall presently need. I return with the book, having forgotten the pamphlet. Then, though I may have no immediate need of it, I am obscurely aware of an unsatisfied impulse to fetch something; I cannot define either the object or its location. That is, I

[1] Note 6.

[2] Writing of planful Activity', he says: 'Within us, if anywhere, we must feel the urge, or however it may be named, which shall afford the basis upon which acknowledgment of Activity is founded.'

suppose, as near as we ever come to experiencing 'blind impulse'. But it is not wholly blind: there is reference to some object undefined; there is reference to place and future time, also undefined; there is even reference to causal efficacy, for that undefined object I am impelled to fetch is thought of as one that will satisfy some undefined need or desire. In such experience we have conation at its simplest. We see in such instances how intimately reference and impulse are bound up together as two aspects of one event. Without reference there could be no impulse; and without impulse or urge there could be no reference to something remote in space or time from the immediately given or present sentience.

If, then, Lloyd Morgan's account of the genesis of 'reference' were otherwise acceptable, we should still have to complain that it neglects one all-important aspect of cognitive apprehension present in every concrete instance, namely, the conative aspect. His account is mechanistic (in our sense of the word); and there would seem to be no possibility that any teleological event can emerge from any conjunction of mechanistic events, even if various conjunctions of sense-qualities be admitted as emergent and synthetically combining.

PLEASURE AND PAIN

The conative aspect of mental events is not the only one neglected by Lloyd Morgan's account of the emergence of Mind. It is generally agreed that feeling (of which pleasure and pain are the simple forms) is an equally fundamental aspect of mental events. And the causal efficacy of pleasure and pain is undeniable, whether we take the hedonist view that they are the essential motive powers of all teleological action, or the better founded hormic view, namely, that their function is to guide or modify the working of impulse, pleasure supporting and confirming it along any given line, pain checking and diverting it. Lloyd Morgan seems to accept the former view. He tells us: 'I believe that conscious guidance does count for progress, and that *au fond* all conscious guidance at the cognitive level is towards pleasure and away from pain or discomfort.' Further, as prospective reference emerges, there emerges also 'foretaste in enjoyment'; and 'if there be what I call foretaste in enjoyment coupled with cognitive prevision of coming events, we have an effective factor in guidance of no little importance—nay more, as I think, of the greatest importance.'

We are told that 'there is no life without enjoyment', that 'enjoyment is concomitant with life' and that 'even so simple an animal as an amoeba has something, however rudimentary, of the nature of enjoyment, something, however incipient, of the nature of reference to its environment. Thus far we do suppose that where there is life there is also mind, though it may be a very simple form of mind.'[1]

But we are told nothing of the emergence of enjoyment or the differentiation of pleasure and

[1] From his essay, 'Mind in Evolution' in the volume, *Creation by Evolution* (London, 1926).

pain from primitive enjoyment; and much of his chapter on pleasure and pain is devoted to a very unsatisfactory attempt to rebut an argument of mine by which I sought many years ago to show that pleasure and pain must be admitted to influence our behaviour in the ways they seem to do, to count in conscious guidance, to make a difference to the course of events.[1]

THE CAUSES OF ORGANIC EVOLUTION

It is significant that the emergent evolutionists say little or nothing of the role of emerging Mind in the evolutionary process. There prevails at the present juncture a tendency to divert public attention from the problem of the genesis of new species and the causes of evolution, and to concentrate upon the task of proving to the public that evolution has occurred. This is, I think, due to the conjoint influence of two features of the present juncture. First, the public or a large part of it has shown a strong tendency to react against the scientists, to refuse to accept evolution lying down, to ask for demonstration, rather than to be content with *ex cathedra* utterances. Secondly, the men of science are no longer so confident and united as they were a generation ago in replying to the question: What caused evolution? They are aware that they have no answer consistent with Modern Materialism and its purely mechanistic principles. A few still claim to find a complete answer in Darwin's great principle of natural selection working upon innumer-

[1] Cp. Note 13, *On Consonance of Pleasure and Welfare.*

able minute variations ; but these few are a laggard remnant.[1]

For most biologists the neo-Darwinians' mutations have taken the place of Darwin's variations. But mutations themselves require much explanation. The emergentist, it is true, may say that the occurrence of mutations affords just another illustration of his great principle, that mutations are emergents. But there are two striking facts about mutations ; first, in so many cases they seem to be adaptive ; and if they are to serve the purpose of evolutionary theory, they must be adaptive. Secondly, any particular mutation seems to appear, not as an isolated phenomenon, but as a whole crop of similar mutations in many individuals at one period in the history of a species. A mutation that is not adaptive is a mere monstrosity ; and an adaptive mutation that appears in one or a few individuals only has little or no chance of becoming fixed by natural selection as a racial character.

If we turn to Lloyd Morgan's *Emergent Evolution*, we find no mention of Darwin or of Lamarck, of variation, mutation, or selection. He frankly asks : What makes emergents emerge ? And his answer is : The directive activity of God. It is very difficult to reconcile this with his explicit repudiation of every form of vitalism, of every theory that

[1] A few even claim that natural selection has been the main condition of the alleged evolution of the physical realm ; but, since natural selection presupposes the struggle for existence and reproduction, which is universal in the organic realm and wholly absent from the inorganic, the principle is quite inapplicable, and the claim remains a bare verbal formula without meaning.

postulates in vital and mental activities the causal efficacy of some factor other than physico-chemical events of the body.[1]

We may at least insist that, when science acknowledges God as the ultimate directive Activity, it does not thereby absolve itself from the obligation to discover if possible in what forms such Activity has manifested itself. And in the sphere of organic evolution the question takes the form : Has it been manifested by way of natural selection, alone or together with the production of mutations for selection to work upon ? Or has it worked through evolving and differentiating powers of mental guidance, of intelligent purposive effort, however humble and obscure the early efforts may have been ?

In other words the issue between neo-Darwinism and Lamarckism remains one of crucial importance for the theory of evolution. According to the former, organic evolution has been purely mechanistic (in our sense of the word) however emergent. According to the latter, Mind or teleological activity has been throughout a guiding influence without which there could have been no evolution.

Now, there are two good reasons for accepting the Lamarckian principle of the inheritance of acquired modifications. First, neo-Darwinism, the theory that rejects it, will not work, not even when combined with the principle of emergence. Secondly, there is a large and constantly increasing mass of

[1] 'The advent of novelty of any kind is loyally to be accepted wherever it is found without invoking any extra-natural Power (Force, Entelechy, Elan, or God) through the efficient Activity of which the observed facts may be explained.'

more or less direct evidence of Lamarckian transmission.[1]

Accepting, then, Lamarckian transmission as a well-based theory, we ask how that bears upon our central problem. The answer is : It bears upon it in two ways, both of the first importance. First, it assigns to Mind a leading role in the drama of organic evolution, instead of regarding it as a mere spectator on the side lines. Secondly, it bears out the conclusion reached in our lecture on memory, namely, that the organization of a living organism is not wholly manifested to us as a spatially distributed material structure, but comprises organization of an immaterial nature. It bears out this conclusion in the following way : If the fundamental assumption of Biological Materialism were true, namely, the assumption that all vital organization is material, Lamarckian transmission could not occur : for it is impossible to imagine any material mechanism through which any modification of form or function acquired by the adult organism can impress itself upon the germ-plasm as a specific modification of it, i.e. a modification of it which will determine the appearance in the offspring of the modification acquired by the efforts of the parent. It is in fact just this acknowledged impossibility that has been the main ground of the great reluctance of so many biologists to accept the Lamarckian principle.[2]

[1] I venture to claim my own prolonged experimental research directed to this question, as affording positive evidence of this kind, perhaps more conclusive than any other. See 'An Experiment for the Testing of the Lamarckian Hypothesis', *British Journal of Psychology*, April, 1926.

[2] I cannot find that the heroic efforts of Dr. E. Rignano to

If, then, Lamarckian transmission occurs, it is in itself good evidence of the reality of that immaterial organization which is indicated by all impartial consideration of the facts of memory, the facts of the unity of consciousness, the facts of integration and disintegration of personality, the facts of intelligent purposive activity.

Let me now try to sum up the bearings of our discussion on the theory of Emergent Evolution.

The principle of creative synthesis seems to be true of Mind or of mental events, and the term 'emergent' may conveniently be applied to the products of such synthesis. But there has been no emergent evolution in the physical realm; and indeed there is no good reason to believe that the physical world in general, or any part of it, has gone through any process of change in some one direction that could properly be called evolution. There has been, not evolution of Mind from the physical realm, but evolution of mental capacities; and this evolution has been characterized by a progressive differentiation of the powers of Mind, rather than by emergence of new kinds of relation, causal or other. Mind everywhere at all levels is teleological, cognitive, conative and affective. New ways of conceiving things and their relations, new theories, new works of art, new inventions, moral choices under novel conditions, all these are instances of emergence, of creative synthesis. But all such events occur according to fundamental laws of Mind

imagine such a mechanism have been successful; and I know of no biologist who does find them successful. Yet his attempt is by far the most thorough that has been made.

that are teleological. There has been no emergence of the teleological from a mechanistic physical realm.

I cannot characterize the scheme of emergent evolution better than by citing the words of a recent critic of it :[1] 'Emergent Evolution appears to be the result of an attempt to find some middle path between mechanism and teleology. It is at least certain that the exponents of this view agree in repudiating even the immanent teleology of Bergson's *élan vital*, and at the same time reject the old-fashioned mechanism of the materialist theory. But it may well be questioned whether this hybrid concept is not destined, like many hybrids, to be sterile. . . . I venture to suggest that it will be found ultimately that the theory of emergent evolution was a convenient halting-place in the passage to a more explicitly teleological conception of nature.'

Whether ultimately it will be found that in the physical sciences mechanistic explanations must give place to teleological only the future can determine. But, pending that unforeseeable decision, we are fully warranted in holding fast to the reality of teleological causation in the mental realm ; and, so long as the physical sciences remain mechanistic, we must be content with a provisional dualism, however repugnant it may be to our aesthetic preferences.

[1] Dr. W. R. Mathews in his Essay ' Philosophy ' in the volume *The Mind* (London, 1926).

CONCLUSION

IN these lectures I have sought to show that, however the physical sciences may conceive the things and events of the physical realm, whatever undefinables and laws they may choose to regard as ultimate or necessary postulates, we are justified in holding fast to our belief in the causal efficacy of our mental or teleological activities, and in believing that teleological causation pervades the organic realm and has been an essential factor of organic evolution.

This view involves certain difficulties which must be frankly admitted, raises certain great problems, the solution of which remains a task for the future. But the fact that such problems remain insoluble at the present time is no ground for flying in the face of the evidence and for forcing ourselves to pretend to be content with the inconsistent and untenable assumptions of Modern Materialism. Every advance of science raises new problems. When the ether was discovered or invented, it was not rejected on the ground that its relations to matter remained (and remain) in many respects obscure and itself a bundle of mysteries or problems. We ought not to expect to be able to solve the problems of Science as soon as they come into view. The mere formulation of a problem in a useful manner may be a long and difficult task, to be followed by a long course of co-operative research. And the difficult problems involved in the acceptance of teleological causation in the organic realm have hitherto hardly been formulated, chiefly by reason of the continued influence of a defunct dogma, that of Atomic Materialism.

It seems worth while to attempt to define the more important of these problems and to indicate however vaguely the lines along which some of them may eventually be solved.

We have seen that the acceptance of teleology commits us to the view that the organization of living things is not purely material organization. The prejudice against any such view seems to be a relic of Atomic Materialism. We are familiar with organizations of many kinds that are essentially teleological in their history and functioning and immaterial in nature. Such are, for example, the organization of a play, of a sonata, of a poem, of a train of argument, of a political party, of a church. It is true that such organizations commonly find expression through and embodiment in material structures, in printed words, musical notations and instruments, written constitutions, buildings and instruments of all sorts that subserve the purpose of the organization. But in this respect animal organisms are in the same case. Can anyone deny that a highly trained orchestra is a teleological organization, which, although it can manifest its nature only through material instruments, is yet entirely dependent for its genesis and continued existence upon a governing purpose? I suggest that such an organization is truly analogous to the organization of all living organisms. The biological materialist takes a piece of living tissue from an animal, displays it in a test tube still manifesting vital activities, and asks : Where is the immaterial organization of the animal that you postulate ? Just so you might take one or two or three players from an orchestra and show them fiddling harmoniously, and might regard that as evidence that the orchestra as a whole is *nothing but* the material masses embodied in it. Or he points out that the animal body contains no material elements not found in inorganic nature ; and asks : Where, then, is the immaterial factor in that organization ? Equally truly the orchestra contains only such material elements as are found in the inorganic realm ; yet these alone are but a subordinate part of the total organization.

The analogy extends to the facts of propagation. A single cell of the animal may transmit and perpetuate the organization of the whole animal ; and a single unit of the orchestra may transmit its organization, perhaps with variations or mutations that give scope for improvement of organization through natural selection, a true emergent evolution.

This analogy points the way to the solution of another great difficulty that is widely felt to stand in the way of a teleological view of such processes as morphogenesis. It is said : We admit that the conscious activities of men seem to be teleological ; but no man consciously directs the growth and working of his bodily organs ; such processes seem to go on quite unconsciously. Since conscious guidance is the only kind of seemingly teleological causation with which we are directly acquainted, how can we legitimately regard these unconscious processes of growth and regulation as of like nature with our conscious activities ? Are they not good evidence that the seemingly teleological conscious processes are really mechanistic ? In other words, if we admit the teleological nature of conscious events, we still have on our hands all the wealth of organic processes which, though they may seem in certain respects to be teleological, do not in any direct way reveal themselves as conscious or mental.

The position, then, is that we recognize a realm of teleological mental events, and a realm of physical events that seem to be purely mechanistic ; and in between these two realms in an uncertain status are all the organic processes that are not obviously mental. Shall these be assimilated to the mechanistic physical realm or to the teleological mental realm ?

It seems to me that the grounds for assimilating them to the mental or teleological realm are overwhelmingly strong. And the principal difficulty in the way of such assimilation may be greatly lessened by the aid of a further analogy. Suppose that in the orchestra, which we have likened to an animal, every member makes the instrument on which he plays. Or change the analogy and compare the construction of a bone with that of a honeycomb. No one of the bees that take part in the construction of the comb (a complicated structure that is nicely adapted to the form and size of the cavity within which it grows, and which is repaired and strengthened according to exigencies arising from time to time, just as in the case of a bone), no one of the bees consciously forms a plan of the whole structure and directs the building operations as a whole. But each bee is a teleological agent exercising a lowly form of intelligence and purpose ; and the natures of all the bees are so nicely adjusted to one another that their joint efforts bring about, in a way we do not fully understand, the adaptations of the growing structure to all the special circumstances that obtain and arise in the course of the work. That analogy gives us, I think, the clue to the way in which we may profitably seek to solve the problems of organic growth and regulation as instances of truly teleological causation.

APPENDIX

Fourteen notes, of which one (Note 6) is an integral part of the argument of the text. The other notes discuss more fully certain topics which are treated very briefly in the lectures.

Note 1. *On the Meaning of 'Mechanistic'*

So long as Atomic Materialism was accepted the meaning of the word 'mechanistic', as applied to qualify any description or explanation, any hypothesis or theory or event, was clear and simple; mechanistic was synonymous with mechanical, and the mechanical theory assumed that all changes were changes of position, all processes were movements of matter, and all causation was of the nature of change of momentum by impact of masses. With the passing of Atomic Materialism it has become very difficult to define in any positive terms the meaning of 'mechanistic', and the negative definition proposed in the text remains, and perhaps always will remain, the only satisfactory and comprehensive one. It is comprehensive and satisfactory; for, owing to the superior certainty and insight that we have of mental processes, we are able to define teleological events clearly and confidently by pointing to instances within the experience of each of us in which the essential features are distinctly recognizable, namely, the foresight of a goal and the urge or impulse or desire for its attainment, the progress towards it, accompanied by some degree of satisfaction and generally, though not in all cases, by some intelligent adaptation to particular circumstances as they arise. We can, then, define mechanistic events by saying they are such as have nothing of these characters. It is true that in respect of many events that we incline to regard as teleological we can obtain no direct evidence of these primary marks of the teleological; and have to judge of their nature by external and less sure indications. But that is a difficulty of practical application that does not vitiate the definition.

This negative definition of 'mechanistic', though not perhaps generally and explicitly accepted, is nevertheless in accordance with traditional usage. Dr. L. T. Hobhouse writes: 'Usage seems, in philosophical nomenclature, to have assigned the term mechanism for the category of explanation from which purpose is excluded.'[1]

The mechanistic, then, as so defined, includes not only strictly mechanical events (if any such there be) but all that can be adequately described and explained without taking account of prospective reference.

Let us notice at once that machines and their workings occupy a peculiar position. The construction of a machine is a typically teleological process, as also its regulation and repair. But, when it works automatically and produces the result for the sake of which it was designed and constructed, it works mechanically, although its working is a link in a chain of teleologically caused events; it is a typical instance of co-operation of mechanistic and teleological causation. Hence, as Dr. Hobhouse truly says: 'The full explanation of our piece of mechanism (a machine) then must include both the analysis of its own operation and a statement of the teleological system in which it forms a part.'

Another way of definining the teleological which seems to me far less satisfactory is that followed by Mr. Broad.[2] Instead of seeking to define teleological causation, he defines a teleological system as one that 'acts as if it were designed for a purpose'. And more at length: 'Suppose that a system is composed of such parts arranged in such ways as might have been expected if it had been constructed by an intelligent being to fulfil a certain purpose which he had in mind. And suppose that, when we investigate the system more carefully under the guidance of this hypothesis, we discover hitherto unnoticed parts, and that these are still found to accord with the hypothesis. Then I should call this system "teleological".'

This attempt at definition of the teleological errs, I suggest, in that it takes for granted just that which we are seeking to define, namely, teleological causation, intelligent purposive action. Under this definition machines become typical instances of teleological systems; and the failure of the definition to define is clear in that, as the author himself points out, a machine is both mechanistic and (in the sense of the proposed definition) teleological. He goes

[1] *Development and Purpose* (London, 1913).
[2] In *Mind and Its Place in Nature* (London, 1925).

on to say that in the same sense a living organism is both mechanistic and teleological, and that, since we do not know positively that any mind designed and constructed the organism, we cannot be sure of its teleological origin. And he deepens the doubt thus cast on the teleological nature of organisms by asserting erroneously that 'many machines are themselves made by machines'. I venture to suggest that no machine has ever made, nor ever could make, a machine, certainly not one like itself, and that this is one respect in which machines differ fundamentally from organisms.

In respect of organisms there are two distinct questions in dispute: first, have they been produced mechanistically or, like machines, teleologically or under teleological guidance? Secondly, however organisms may have been produced, are the events within them purely mechanistic, or are they teleologically guided?

Another proposed definition of the mechanistic, at present widely in favour, especially with the exponents of Emergent Evolution [1] is that mechanistic events are in principle predictable and all non-predictable events are non-mechanistic. This way of defining the mechanistic is a curiosity of contemporary thought which can hardly be maintained. It has been engendered by the desire for principles of explanation of organic events less inadequate than the mechanistic, crossed by the aversion from the recognition of truly teleological causation; it is, I suggest, a sterile hybrid. Let us examine it more closely as propounded by the exponents of Emergent Evolution. Lloyd Morgan writes: 'It is pretty certain that the interpretation of nature I put forward will, in some quarters, be characterized as mechanical and vitiated throughout by an uncritical acceptance of what is sometimes spoken of as "the mechanistic dogma". The odd thing here is that the whole doctrine of emergence is a continued protest against mechanical interpretation, and the very antithesis to one that is mechanistic. It does not interpret life in terms of physics and chemistry. It does not interpret Mind in terms of receptor-patterns and neurone routes. Those who suppose that it does so wholly misapprehend its purport. . . . The essential feature of a mechanical or, if it be preferred, mechanistic interpretation, is that it is in

[1] The doctrine of Emergent Evolution is examined in chapters V and VI.

terms of resultant effects only, calculable by algebraical summation. It ignores the something more that must be accepted as emergent. It regards a chemical compound as only a more complex mechanical mixture, without any new kind of relatedness of its constituents. It regards life as a regrouping of physico-chemical events with no new kind of relatedness expressed in an integration which seems, on the evidence, to mark a new departure in the passage of natural events. Against such a mechanical interpretation —such a mechanistic dogma—emergent evolution rises in protest. The gist of its contention is that such an interpretation is quite inadequate. Resultants these are; but there is emergence also. . . . That it cannot be mechanically interpreted in terms of resultants only, is just that for which it is our aim to contend with reiterated emphasis.' [1]

This proposed distinction between the mechanistic and the non-mechanistic is really one between the strictly mechanical explanations of Atomic Materialism and the wider kind of mechanistic explanation that recognizes the inadequacy of strictly mechanical explanations to many inanimate events, especially chemical events, and asserts that chemical compounds exhibit properties that cannot be deduced or predicted from a knowledge gained by studying in isolation the elements that enter into them. And, since living organisms are the seat of many complex chemical processes, it follows that their processes are, in the sense proposed, unpredictable and therefore non-mechanistic. The term 'emergent vitalism' has been proposed to denote a view of organisms that is non-mechanistic in this sense. But the expression 'emergent vitalism' is hardly appropriate; for, according to this view, there is no essential distinction between vital processes and those dealt with by all chemistry. And indeed both this criterion of the mechanistic as the predictable and the term 'emergent vitalism' are proposed for the purpose of exhibiting vital processes as not essentially different from other chemical processes.

I suggest that the proposed distinction between predictable and non-predictable processes is one that holds good within the sphere of mechanistic processes, rather than one between the mechanistic and the non-mechanistic. And we may go further and question the validity or usefulness of this criterion of predictability.

[1] *Emergent Evolution* (London, 1923).

Prediction is in actual practice a matter of foretelling with more or less probability. To predict is to make an inference, to reason to a conclusion. And only those conclusions are logically certain that rest upon conditional premises. Thus, *if* all men are mortal, and *if* Socrates is a man, then Socrates is mortal. That conclusion is logically sound ; *if* the premises are true, Socrates will die. But in the world of nature our premises are merely inductive generalizations. We can surely say : If all oxygen and all hydrogen combine, at a certain temperature, etc., to form water, and if this be O and that be H, then this and that will combine to form H_2O. But we cannot surely make this prediction omitting the conditioning *ifs*. There may be conditions hitherto unknown under which O and H will combine to yield something having other qualities (e.g., H_2O_2). And it may be that this is not oxygen in the sense in which the words are used ; it may be, for example, in spite of the most delicate investigation of it, a hitherto unknown allotropic form of O which will behave in relation with H in some way unpredictable because hitherto unobserved.

Even in dealing with teleological events prediction with a high degree of probability is possible. We can predict statistically ; as, for example, concerning the number of deaths from suicide in a given population. Or we can place a hungry child or dog in the presence of appetizing food and can predict with very high probability that he will eat some of it.

Predictability, then, is practically quite unsuited to be the criterion of the mechanistic. And as a practical criterion there is another grave objection to it. Emergent Evolution regards the qualities and properties of all chemical compounds as emergents ; that is to say, it regards them as unpredictable prior to observation of them, and therefore as non-mechanistic. But Emergent Evolution regards also the qualities and properties of the atoms as emergents ; they also could not have been predicted before they had emerged : it follows that all physical processes in which atoms play a part (even those that do not involve chemical change) are also non-mechanistic. Even the communication of motion by impact of one mass of matter upon another involves the emergent properties of atoms, and is therefore non-mechanistic. In fact, no events are mechanistic, except those that involve no ' emergents ', those that occurred or occur in the hypothetical primordial world-stuff that has undergone no emergent evolution. Hence all the events which we can in any way observe are equally unpredictable and non-mechanistic until after we have observed events of a similar kind, and the distinction loses all value and significance.

Again, there is serious doubt about the unpredictability of chemical events. No one questions that after we have become acquainted with the chemical substances concerned, we can predict many such events with high probability, say the formation of water, or of sodium chloride. It is alleged only that such chemical events are unpredictable in the light of knowledge confined to the qualities and properties of the elements before they were combined, before the properties of H_2O or of NaCl had been studied. Now it may be that no physical-chemist could have predicted all the properties of H_2O before having studied it; and yet it seems probable that this incapacity to predict is only an expression of ignorance of the nature of H and O. If on their combination H and O yield water, presumably they contain in some sense the potentiality of forming water. In fact it is of the essence of Emergent Evolution that nothing new is added from without, that 'emergence' is the consequence of new kinds of relatedness between existents. The presumption is, then, that with sufficient knowledge of the components, H and O, and of the general principles of chemical combination, highly probable predictions of the properties of water could have been made. In fact chemists have successfully predicted the properties of compounds they have never observed and have proceeded to produce those 'emergents'. They have even predicted the existence and the properties of elements which had not been observed.

I fancy that most physical-chemists would claim that it is the programme of their science, a programme already achieved in part, to render all chemical events predictable, even those that result in compounds having properties not previously observed.

I have argued above that, according to the principles of Emergent Evolution, all events open to our observation are equally unpredictable and therefore equally non-mechanistic; for they all involve so-called emergent properties or qualities. It may be answered that it is only those events in which emergence of new qualities actually occurs that are to be regarded as unpredictable and non-mechanistic ; while events in which emergents take part, but in which no new emergents appear, are mechanistic because predictable in the light of our empirical knowledge of the emergents concerned. This, I think, is what the emergentists mean to maintain. But, if this is their contention, it follows clearly that their criterion of the non-mechanistic is wholly relative to the state of science at any moment. Emergent qualities, it is said, cannot be predicted ; they can be known only after they have emerged. But each emergent process becomes empirically predictable after it has been observed in a sufficient number of instances. It would follow that every event of a kind we can observe was originally non-mechanistic and so continued until science had formulated an empirical generalization that enabled the prediction of similar events ; thereupon events of that kind became mechanistic. This is true even of strictly mechanical events, that is those that can be explained and predicted on the basis of Newtonian mechanics ; for the laws of mechanics are also empirical generalizations, and, until they had been formulated, such events remained unpredictable. Newton's first law of motion is an empirical generalization, and so is the law of the uniformity of nature. The flight of an arrow was, then, an emergent and non-mechanistic event up to the date of the formulation of the laws of mechanics ; and since that date it has been a mechanistic event. In short, according to the principles of Emergent Evolution, all events that cannot be predicted and explained by deduction from empirical generalizations already formulated, all such events are 'emergent' and non-mechanistic ; but they become mechanistic as soon as the necessary empirical generalizations have been made. Thus the proposed distinction between mechanistic and non-mechanistic is one that has regard not to the nature of things but only to our knowledge of them.

Mr. Broad suggests that each chemical combination is the expression of a unique and ultimate law and therefore is an emergent event (*Mind and Its Place in Nature*, p. 65). It would follow according to his principle that every process of chemical combination is an emergent non-mechanistic event, while processes of chemical decomposition are mechanistic. But it is not true that each instance of chemical combination expresses 'a unique and ultimate law'. If silver chloride had never yet been observed, chemists could predict with high probability many or all of the qualities and properties which it is actually known to exhibit, for example, its weight, all those properties which characterize the class of metallic salts, its non-inflammability, its indifference to many chemical agents, its liability to be acted on by others, its volatization at sufficiently high temperature, etc., etc. It is, I take it, impossible to assign a limit to the power of prediction or explanation which chemical science may attain in such matters.

It has been suggested that the atoms of the various chemical elements are 'planetary' systems of protons and electrons, those of the various elements differing from those of other elements in the number and arrangement of these sub-elements. Suppose this view to prove itself increasingly useful and valid in the interpretation of chemical phenomena. Is it credible that increasing knowledge of this sort should throw no light on the qualities and properties of chemical compounds, should leave them all 'emergents', should not reduce them more and more completely to the status of resultants. Is it not clear in the light of this possibility that predictability means explicability in terms of empirical generalizations of causal efficacy, that the event which is more or less predictable is correspondingly explicable in the sense that it can be regarded as an instance of a class of events already known ?

There is yet another way of examining the question, and the whole matter is so important that I may perhaps be excused for labouring at it further.

It is said that the combination of two chemical elements (say Na and Cl) involves 'emergence', because 'the law connecting the properties of silver chloride with those of silver and of chlorine and with the structure of the compound is, so far as we know, a *unique* and *ultimate* law'. (Broad, op. cit., p. 65). This, I say, is certainly not true of all the properties of NaCl. But may it be true of any one quality or property ?

It is admitted, it is asserted, that the alleged emergent quality or property of NaCl is an expression of qualities or properties, or of some quality or property, hitherto latent in the elements Na and Cl or in one of them. Suppose it to be the expression of a property hitherto latent in Na. Now the assertion that the emergent was unpredictable in principle implies that this hitherto latent property of Na (call it X) could not have been revealed, could not have entered into effective relatedness, could not have played its causal part in any other events, in any other circumstances than those in which it enters into combination with Cl. I submit that this proposition has little plausibility; that, on the contrary, it is highly probable that an exhaustive study of Na in all its other chemical reactions, a full knowledge of its place in the periodic table of the elements and of the properties of such substances as NaBr (perhaps a detailed knowledge of the structure of the Na atom, the number and orbits of its electrons and so forth) would throw light upon all the qualities and properties of Na, including the alleged latent quality revealed on the 'emergence' of NaCl.

Summing up on predictability as the criterion of the mechanistic, we may say: The emergentists' view implies that some events, those that are called mechanistic, can be predicted or explained from *a priori* principles, while non-mechanistic or 'emergent' events are such as can be predicted or explained only in the light of empirical generalizations. But all physical events are of the latter kind; therefore there are, under this view, no mechanistic events.

The common view of teleological causation is that it is a non-natural or miraculous interference with the course of mechanistic events, some kind of influence of which nothing more can be said, one of which no laws or generalizations can be formulated. Under this view,[1] for which there is no justification, though its historical origins are obvious enough, the postulation of teleological causation is the negation of all science, of all attempt to understand. The prevalence of this naïve view of teleology accounts for the general scorn of men of science and of most philosophers for teleological explanations. It is really a survival from Atomic Materialism; for the acceptance of the latter involved the assumption that all causation is perfectly transparent and intelligible in terms of the simple ultimates of mass, motion and momentum. This common attitude towards teleology is also confusedly connected in the minds of those who hold it with the principle of strict determinism, held either as a necessary assumption of scientific investigation and explanation or as a necessity imposed upon us by

[1] Pretty clearly expressed, for example, by Professor Durant Drake in his *Mind and Its Place in Nature.*

the constitution of our minds. For it is felt by them that to admit the reality of teleological causation would be to admit in the course of events irregularities not conformable to laws or generalizations of any kind.

It seems worth while to ask : What is the relation of the principle of determinism to mechanistic explanation and predictability ? Predictability of events is commonly equated with their deterministic nature. It must be admitted that any realm of strictly predictable events must be deterministic. But, as we have seen, no events are strictly predictable ; we can predict in any case only with more or less of probability on the basis of inductive generalizations. The wider our basis of induction and the more the particular event predicted is connected by empirical laws with other events similarly generalized in laws, the more probable becomes our prediction. Such illegitimately assumed strict predictability of all events seems to be the only ground of the assumption of universal determinism and of the denial of human freedom to choose or create.

Nor can the realm of determinism be equated with that of mechanistic causation. If all events of the physical realm are strictly mechanistic, it may nevertheless be true that in the course of those events occur points of strict indetermination, conjunctions at which the causal factors may produce equally well either one of two alternative courses. On the other hand it may be that teleological events are strictly deterministic. We cannot positively assert that they are not ; though we can say that such a view seems very improbable.

A determinism that knows its business, understands the principles of science and the working of our minds, will then be content to say that the assumption of determinism has worked well in physical science, and continues to be useful ; and that we may with advantage make provisional use of the same assumption in the biological sciences ; that is to say, we may use determinism, not dogmatically as an established principle, but as a working hypothesis, one to be abandoned where we find sufficient reasons for so doing.

NOTE 2. *Differentiation of Science and Philosophy*

The distinction drawn in the text between Science and Philosophy is not yet orthodox doctrine, and I suppose that many philosophers would not accept it. Yet the history of the development of science and philosophy seems clearly to indicate that this is the true distinction and that the differentiation of their provinces which has slowly been taking place can end only with the general acceptance of this demarcation of their respective functions.

Some philosophers will refuse to accept the distinction because it will seem to them to belittle the functions and to restrict unduly the field of philosophy. Yet the acceptance of it will leave to philosophy important rights and duties in every sphere of human thought. Philosophy may call upon every branch of science to examine its presuppositions or postulates, to revise its procedures and harmonize its conclusions with those of other branches. The field of philosophy will still be co-extensive with the field of science and even more extensive. There will still be room for and indeed need for a philosophy of science or of each of the sciences, the physical, the biological and the mental, philosophy of physics, of biology and of Mind. Beside the science of human behaviour we shall need the philosophy of conduct; beside the science of aesthetic, the philosophy of the beautiful; beside political science we shall need political philosophy; beside the science of religion or the religions, a philosophy of religion.

If philosophers would consent to recognize and would strictly observe the proposed demarcation of their functions they would avoid provoking, as at present they do not always avoid, the scornful antagonism of men of science; they would find themselves in a position of more secure dignity and no less usefulness than they have occupied in the past; especially would this be true, if they also recognized, as not all of them have done in the past, that while they must abstain from trespassing upon the field of science, from presuming to assume its functions, they yet need, in order effectively to occupy their own field, in order to exercise their own functions, the richest possible acquaintance with the content of the sciences.

NOTE 3. *Sensationism and Neo-Realism*

Science, or modern science that knows its business, does not commit itself to any statements as to what substance or substances the world is made of. It does not necessarily conceive the physical world as spatially extended or tri-dimensional. It admits that Kant may have been right in asserting that space is a form which our minds, by their very constitution, impose upon the phenomena or appearances to us of the physical world. But it cannot give up the view that in sense-perception we do apprehend, however distortedly, something of the nature of things other than ourselves. And it cannot give up as false or ill-founded our temporal way of conceiving events, i.e. it holds that events exhibit duration, sequence and succession, that change goes on. Further, Science cannot give up the principle of causation, the view that events are bound up in a web of reciprocal influence.

It was from the scientific point of view the gravest defect of absolutist metaphysic that it reduced not only space but also time and causation to mere appearance, while it was as destructive of all human values as the most rigid Atomic Materialism.

Some philosopher-scientists, notably the late Ernst Mach and Professor Karl Pearson, rejecting the thing-in-itself, have affirmed that we know and can know only sensations ; that we should be content to conceive what is called the physical world as consisting of sensations.

This proposal, though it retains time as real, implies the repudiation of objective space and causation. It would replace causation by mere correlation in time. It is thus closely connected with, is in fact a special form of, the view that the task of science is, not to explain the course of events, but merely to render a description of phenomena as full and accurate as possible in general terms. If this view of the function of science is well-founded, our inquiry falls to the ground, *caedit questio.* For the question which we are seeking to answer, the question of the validity of Modern Materialism, is essentially the question : Is the mechanistic type of causal explanation the only valid one ?

Is not explanation by teleological causation also valid, either within the realm of Mind or of organic nature, or possibly throughout all nature ?

We must, then, examine this view which would make of our inquiry nothing more than a pursuit of a will-o'-the-wisp, an endeavour to answer a question that should never have been asked. First, History shows that the progress of Science has been very largely due to the formulation of ever new hypotheses for the causal explanation of the phenomena we observe and to experiments directed to the proving of such hypotheses. The hypothesis of causation by the communication of motion and transmission of momentum is one such. Others that have greatly contributed to the advance of science are the explanation of gravitation by the law of attraction ; of chemical action by chemical affinity ; of light by ether-waves or by corpuscular emission ; of the origin of species by natural selection ; of heredity by the transmission of particles in the germ-plasm. The list might be greatly extended. When any such hypothesis ceases to be useful and used, it is not that Science is content to leave the phenomena in question unexplained ; it is because the hypothesis has been superseded by some one more useful. There is no good reason to believe that in this respect Science will undergo any radical change of method.

Secondly, the utter untenability of all views that would reject causation (from Hume to Pearson) appears very clearly if we consider the purpose of Science and the motives of scientific endeavour. The purpose of Science is not merely to know or faithfully to represent or describe nature, it aims also to control natural events ; and, though sheer curiosity, the desire to know and to understand, has been an important motive, the desire to foresee the course of events, and, foreseeing, to intervene and control that course, for the sake of human welfare, this desire in every age has been a more powerful and fruitful motive of inquiry into the nature of things. Now to control, to intervene, to modify the course of events, is to make practical application of the principles of causation. Take the field of Eugenics with which Professor Karl Pearson is particularly concerned. His statistical correlations are of great value. But Eugenics aims to maintain or improve the natural qualities of human populations ; and, when we seek to apply it with this purpose, we have to devise means of working upon human motives, we have to recognize them as causes that may be influenced by knowledge and by various social conditions and changes.

In the main the superiority of modern to ancient science lies, not in the superior accuracy and fullness of its descriptions, but rather in the consistency of the system of causal explanations which it offers. The Ptolemaic description of the heavens might be as accurate as the Copernican. The circulation of the blood might have been accurately described in full detail; but if it had been explained by attributing to the venous blood a desire to reach and absorb the oxygen in the lungs, and to the arterial blood a desire to deposit its load of oxygen in the muscles and other tissues, our understanding and power of control over the circulation would have been less effective than it has become through the system of mechanical explanation introduced by Harvey. The description of the animal and vegetable kingdoms and of the affinities and relations of descent between species and genera might have become as detailed and accurate as it now is without any hypothesis for the causal explanation of the origin of species; but without such hypothesis progress would have been slower and we should not have learnt how to produce by artificial selection new and useful varieties.

The sensationist doctrine which we have just now examined and rejected may be called a form of idealism; yet it has affinities with another modern doctrine known as Neo-Realism; and the lay reader may be in danger of confusing the two. Neo-Realism teaches that the qualities of physical objects which we seem to become aware of in sense-perception, their colours, tones, odours, tastes, warmth and coolness, really are, as common sense has usually assumed, properties of the physical objects. This involves a going back upon a distinction which seems to have been first clearly drawn by Galileo (though it goes back to Aristotle and is commonly associated with the name of John Locke) and which has become pretty well established in modern science; the distinction, namely, between primary and secondary qualities of physical objects. It seems that the spatial properties of physical things, their shapes, sizes, positions, motions and masses, belong to them in some more real sense than do other qualities, such as colour and odour, which we seem to discover on perceiving them. Colour is commonly explained by Science as a peculiar effect upon the perceiving mind of the combination of light-rays reflected from the object into the observer's eye; the odour as a similar effect produced by chemical particles floating off from the object and affecting the olfactory nerves of the observer. And similarly of the other secondary qualities. Now it is true that we may doubt whether the spatial properties and relations of things are actually such as science describes them to be; and we may, with Kant, hold that the way we conceive them depends at least as much on the constitution of our minds as on that of the things we perceive. Consider a red ball. Its colour as we perceive it is an effect upon our consciousness of the ball's peculiarly selective reflection of light rays; but its shape, as I perceive it, is also an effect of the peculiar way the light-rays reflected from it are grouped upon my retina. If I interpose a blue glass between it and the eye, its colour seems to be other than red; and if I interpose a distorting lens, its shape appears to be other than spherical. Or again if my colour-vision-apparatus is disturbed, I no longer see the ball as red; and if the refractive media of my eye are disturbed, I no longer see it as spherical. It is, then, not only on account of superior stability in perception of the spatial qualities, nor because, as Aristotle insisted, they are perceptible by more than one sense, that this distinction between primary and secondary qualities has been widely accepted by science. It is rather because the spatial properties of things are those which we invoke in all mechanical explanation. The principle has never been more clearly stated than by Galileo. He referred to the tickling we experience if the end of a feather is introduced into the nostril. Shall we say the tickle (a peculiar quality of sensation) is a property of the feather which it somehow communicates to my mind? Or shall we explain the tickle as an effect on me of the mechanical properties of the feather acting via my nostril and its nerves? And we might add: If we burn the feather, is the peculiar odour a property of the feather or of the flame? or is it an effect produced in me when I sniff in particles emitted by the burning feather? Clearly, in both cases the later statement comports much better with, lends itself better to, the system of mechanical explanation of all physical events. And the establishment of this distinction by Galileo was, as a matter of historical fact, an important step in the setting up of Atomic Materialism. For this reason, some modern philosophers [1] regard the acceptance of the usual distinction between primary and secondary qualities as the very essence of materialism. That, however, is going too far. The distinction has a relative validity; and so long as physical science works with mechanistic explanations (which perhaps it will and should continue to do indefinitely) it will find this distinction useful and therefore valid, if not indispensable. For anything we know the mechanistic principles may be strictly applicable to and valid of the physical or inorganic world; and we have no right to deny their validity merely because by so doing we are enabled to construct a cosmology that comports better with our scheme of values.

But there are objections to Neo-Realism more serious than its repudiation of this useful distinction between primary and secondary qualities; namely, it conflicts with a multitude of facts brought to light by the intimate study of the processes involved in sense-perception. Neo-Realism does not, would not, if we could accept it, deprive our inquiry of all meaning and purpose (as sensationism does). But it is necessary to notice it here, because it represents one line along which modern thinkers are striving to overcome Modern Materialism. Its exponents have realized that Modern Materialism has arisen through the too abstract treatment of nature by Science; it represents their endeavour to rectify this treatment by restoring to the scientific and too abstract description of natural objects features which, it holds, properly belong to them and which have been unduly neglected hitherto by Science, neglected just because they seem to contribute little or nothing to the mechanistic explanation of physical events. This endeavour is no doubt praiseworthy; though the form it has taken in the hands of the Neo-Realists seems to be mistaken. But, since the rejection or acceptance of Neo-Realism does not vitally affect our argument, I do not delay to display the insuperable nature of its difficulties. I will only point out that it seems to me to be due to the influence of an ancient psychological error; the error, namely, that what we normally perceive in sense-perception is the sensory qualities of objects. But to this topic we must return in another connexion.

[1] E.g. Professor R. A. F. Hoernlé, in his essay in *The Mind: a Series of Lectures*, edited by R. G. S. McDowall (London and New York, 1927).

NOTE 4. *The Psycho-biological School*

There is a school of biologists who, without investigating the question of the reality of teleological causation, assume the validity and propriety of teleological interpretation of biological events ; leaving open the question of the relation between the biological and the inorganic realms. Leading representatives of this school are Drs. J. S. Haldane and E. S. Russell. The latter (*The Study of Living Things*, London, 1924) describes the method as the ' psycho-biological or functional method which seems to offer the best, if not the only way of escape from the materialistic conception of the living thing '. He writes : ' There are in fact only two legitimate ways of regarding living things : either we may regard them as material simply, or we may make the hypothesis that the reality underlying their material appearance is an activity similar to that which we know to be the inmost reality of our own life. . . . If we adopt the second alternative, we must interpret all organic activities as in some sense the actions of a psycho-physical individual.' Further : ' The essential difference between the inorganic unit and the living individual is that the activities of all living things tend towards some end and are not easily diverted from achieving this end. Let no one say that this is a fanciful reading of the facts. It is impossible for anyone, whatever his philosophical prejudice may be, to deny that all goes on in the organic world *as if* living beings strove actively towards an end, whether of self-development, self-maintenance or the continuance of the race. And if we interrogate our own deepest experience we find, as the ultimate driving forces of our life, deep-seated conative tendencies or instincts, of which we may normally be totally unconscious, but which impel us to courses of action which we may or may not be able later to rationalize. We may conclude then that what differentiates a living thing from all inorganic objects or units is this persistence of striving, this effort towards the expression of deep-lying instinctive tendencies, which we know to be the inmost core of our own being.'

Of biology of this type, he says : ' Beyond its fundamental principle that the living thing is to be regarded as an active and enduring individual, the psycho-biological view implies no theoretical assumptions. It admits of no action of mind upon body, has recourse to no hypothetical immaterial agencies or entelechies. It aims at an objective constation of the responses of the organism in development, behaviour and functional adaptation, and does not attempt to penetrate the inner quality of the organism's experience nor to supply a purely psychological explanation of all its responses.' Again: 'It must, however, be borne in mind that *functions*, in the sense in which the word is here employed, are essentially of the same nature as *responses*, and imply the elementary psychical moments, shown as tendency, mnemic action and regulability, which characterize the responses of the organism as a whole. A cell or organ function, as here understood, is not the materially determined outcome of the physico-chemical configuration —that is a definition which derives from the materialistic philosophy—it is an activity *sui generis* of the living unity, not completely reducible to physico-chemical processes, though dependent upon these.' And of the processes of bodily growth he writes: 'While development depends upon the activities or functions of the developing parts it is too centralized and unified to be regarded otherwise than as a fundamental activity manifested by the whole. The responses or activities which come into play are extraordinarily complex and their interrelations are manifold, but they are themselves ruled, in a way which still remains one of the deepest mysteries of life, by the fundamental "Trieb" or urge towards the attainment of the typical or specific form of the adult whole.'

Now, biology of this type is certainly very superior to the mechanistic biology still widely in favour. And it is noteworthy that working biologists should thus frankly recognize the inadequacy of mechanistic interpretations of biological facts and should insist upon their teleological nature. But, though the assumptions of this school may suffice for the practical purposes of biological research, the human mind will never rest content with setting apart the organic and the physical realms as radically different. We inevitably ask: What are the relations between the two realms? What are the relations between the hormic and the mnemic tendencies which Russell recognizes in all living things and the physico-chemical processes which he acknowledges to be necessary to and indispensable to the physiological events that take place in them? This school seems to ask us to restrain this natural desire to understand, asks us to be content to regard the biological sciences as distinct in method and aim from the physical sciences. But we cannot, nor ought we, to conform with this demand. This setting up of two kinds of science not intelligibly related to one another is, I suggest, a dualism far more radical and vicious than the psycho-physical dualism which Russell arbitrarily rejects on wholly inadequate grounds.

Note 5. *The Role of Meaning in Action*

We have held chiefly in view in the foregoing discussion our appreciation of causal relations as an all-important factor of many forms of intelligent action. We might insist equally forcibly upon other aspects of or modes of intelligence as pointing clearly to the same conclusion. Temporal, spatial and causal relations are meanings we read into the sense-impressions we receive; to think such relations is our way of reacting to the sensory cues. But there are many other forms of meaning that, in similar fashion, we read into or add to various sensory cues.

The extremer mechanists are for ever telling us that our behaviour is a series of responses to physical stimuli. This is one of those statements that are far worse, far more misleading, than a plain untruth: for it is insidious; it has just enough of truth to beguile the simple-minded into acceptance, and enough of error to bar them for ever from further insight into the nature of our mental life.

Let us consider very briefly another great class of intelligent appreciations by which our actions are to a great extent guided, namely, our perceptions of the signs and signals given us by our fellows. When these signs or signals take the form of words, it is very obvious that the nature of the physical stimulus is of no importance for the determination of our action. The words may be spoken or written, whispered or shouted or sung, and they may be words of any one of the several languages that are familiar to you. And your response is in all cases the same. Your response normally depends upon the meaning the words have for you; and, so long as language performs normally its function of communication, the words evoke in your mind the meaning that was in the mind of the speaker, the meaning that he desired to communicate to you and that he expressed in the words uttered. The words then are merely symbols; as sense-impressions they are like other sense-impressions, merely cues that provoke us to think of that which they symbolize.

Words have the advantage of fixation of meaning by convention; but in many respects they are as symbols much inferior to gestures and facial expressions. The young child and the companionable dog, who take the meanings of but few words, share with us the ability to find rich meanings in gestures and facial expressions. When we (or they) watch the face of a friend, we do not perceive merely certain varying patches of colour and of light and shade and shifting lines ; we perceive his emotional states and changes, his desires, intentions and much else, and our reactions are governed by our appreciation of these meanings. Suppose that a man says to you 'Coward!' Your reaction is determined neither by the physical characters of the stimulus, nor by the conventional meaning of the word, but rather by the whole highly complex meaning of the situation, the complex of emotional relations between you. The word may mean for you merely a playful stimulus to greater effort, a serious reproach, or a wanton insult ; and your reaction is determined by the meaning you read into it, by the intention with which the word was uttered and which you perceive.

I have added these remarks to our discussion of our appreciation of temporal, spatial and causal relations in order that we may have clearly in mind the wealth and variety of relations that we apprehend, even in relatively simple instances of perception ; and the fact that it is our appreciation of these relations, rather than any complex of sensations, that guides our actions.[1]

[1] In my *Body and Mind* I have argued that, while the sensory qualities we experience have their immediate correlates and determinants in corresponding brain-processes, the meanings we think have no such brain-correlates. There is at present no possibility of producing conclusive evidence of the truth of this view. But it is worth noting that this view, which was very heterodox at the time it was propounded, has become an accepted implication of the now so fashionable doctrine of Emergent Evolution.

NOTE 6. *Some Attempts to Exhibit Teleological Causation as Crypto-mechanistic*

In connexion with our discussion of the meaning of the word ' mechanistic ', we have noticed certain attempts to set apart some events of the physical realm as non-mechanistic and as, in so far, of the same order as vital events, attempts which, if successful, would abolish as invalid all radical distinction between, on the one hand, vital and mental events and, on the other, these alleged non-mechanistic physical events, such distinction as is involved in regarding the former as teleological, the latter as mechanistic. Similar in their general tendency are various attempts which, recognizing the seemingly teleological nature of mental events, proceed to explain away this appearance, to explain it as a disguised form of mechanistic happening.

One such way of obscuring the essential difference between teleological and mechanistic causation is that propounded by Wilhelm Wundt in the interests of Psycho-physical Parallelism. Wundt proposed to accept this alleged parallelism with ' natural piety ', as the emergentists have it, saying that the difference between a mechanistic series and a teleological series is merely subjective ; that is to say that the mechanistic and the teleological are merely two ways of regarding any train of events ; that if we trace a train of events back to a point in their past history, we regard it as mechanistic ; while if we trace it forwards to a future point, we regard it as teleological. But we cannot save the reality of teleological causation by means of any such easy verbal trick. If one train of events conforms to mechanistic causation, and another to teleological causation, if the former is determined wholly by the past without reference to the future, while in the other reference to the future makes a difference, the two trains cannot be universally parallel. It may be valid to suppose that occasionally different causal sequences lead to similar results ; but to suppose that two so entirely different sequences as the mechanistic and the teleological invariably do so is absurd.

Wundt's proposal, in short, would reduce causation to a purely subjective interpretation of events without objective validity and thus would undermine all science.

Prof. Warren's Study of Purpose

A leading American psychologist, Professor H. C. Warren, has attempted to explain away purposive action, to explain it in terms of the mechanistic association-psychology which he expounds.[1] This seems to be the most thorough and competent of all such attempts and deserves careful examination. The older associationists were content to say that the idea of a goal is formed by previous perception of similar goals, that this idea has become associated with the idea of the movement or train of movement necessary for the attainment of the goal; and that in purposive action we have merely a sequence of associative reproduction of one idea by another; to which is commonly added that, of course, it is the brain-processes correlated with the ideas that really do the trick.

Warren has the merit of having seen that this is a pitifully inadequate story and of trying to improve upon it. His purpose frankly is to give 'an interpretation of purpose in physiological terms, as a series of physico-chemical changes'. 'There is', he says, 'no reason why the most thorough-going mechanist should not accept purposive events as a specific class of natural processes,' i.e. as a special class of mechanistic events. And he fully realizes the momentous nature of his undertaking. We conceive, he rightly says, all teleology or purposive process in the light of our own experience of goal-seeking. All teleology is founded upon that; and, if we can show that the clearest instances of our own purposive action can be mechanistically explained, teleology of any kind has no leg to stand upon and must fall to the ground. 'Science to-day stands face to face with the problem: What is the place of "purpose" in the universe as a whole?' And the answer must turn upon the success or failure of psychological attempts of the kind he undertakes.

He rightly points out that simple actions preceded by no deliberation may 'contain the essential features of purpose and are quite as typical'; and that, though 'Purposive consciousness is observed directly only through personal experience', yet 'its presence in another individual, like the presence of mental data generally, may be inferred from the character of his reaction or from his verbal report'.

[1] 'A Study of Purpose', *The Journal of Philosophy, Psychology and Scientific Method*, vol. XIII, 1916.

He distinguishes five characteristic factors of purposive consciousness, namely: (1) an idea of some future situation; (2) decision, wish, or assent; (3) a feeling of potency; (4) consciousness of self; (5) a feeling-tone of fitness or unfitness.

Of these the first is the most essential. 'Anticipation is the fundamental characteristic of the purposive experience.' How then does an anticipatory idea differ from a bare idea or a retrospective idea? The author has simply not seen that there is any such problem before him. He merely dilates upon the thesis that there is no essential difference between an idea of a goal and an idea of means for its attainment. The answer can be given only in terms of the forward-looking urge which is of the essence of purposive action and which probably is one of the ultimates that we must accept with 'natural piety', because it is not explicable in terms of anything else. But Warren simply overlooks and ignores this fundamental difference between forward-looking and retrospection. If he were pressed he would, I suppose, answer according to the stock formula of associationism: The anticipatory idea is one that is associated with an idea of the future. To which we retort: How comes an idea of the future, if we have never looked forward?

In discussing assent Warren rightly distinguishes between mere imagination of the future and purposive forethought; they differ, he says, in that in the latter is present an additional factor 'the consciousness of intention, decision, or volition—of wish, desire, or assent'. This is an 'affective datum', it is 'a feeling of actually fulfilling the idea'. And if we analyse it, we are told, 'we find that it consists of kinesthetic and organic data' and arises from incipient tension of the muscles. The sensory effect of these incipient tensions is the affective datum which is the assent. It would seem that the fact of desire is somehow included and dealt with in this strange doctrine of assent as kinesthetic sensation; for under this head we are told that 'the function of desire is only to increase the intensity of the forethought and thus render its accomplishment more probable'; and we are told that long-continued 'hard thinking' may intervene between the forethought and its fulfilment. Intensification of forethought (whatever that may mean) and hardness of thinking seem then to be effects of that affective effect of incipient muscular action which is assent. Suppose now we admit that the sensations of incipient muscular action may have this wonderful efficiency in intensifying forethought and producing hard thinking. We have still the fundamental problem on our hands: Why in mere forethought no incipient muscular action, and in purposive forethought the incipient muscular action that makes all the difference? Warren's account amounts to this: Mere forethought is not accompanied by incipient muscular action, purposive forethought is so accompanied. But on the question, why one is so accompanied and the other not, he throws not the slightest light. And yet this is (in his peculiar terminology) just the problem he sets out to solve in terms of mechanistic associationism. Note that if he had answered this question he would be no nearer an explanation of purposive effort; he merely ascribes to muscular sensation, the 'mystic potency' of causing 'hard thinking' and all the phenomena of effort or striving.

Even in the most anxious deliberation issuing in a resolute decision 'the assent element is still the same; it is merely magnified'; i.e. it consists of more intense muscular sensations or more of them.

In the discussion of the third factor, potency-feeling, we learn that this is the dynamic factor in purposive process; our doubts as to the efficiency of muscular sensations in making our thinking hard or intense are therefore intensified. But the substitute offered is hardly more satisfactory: for 'the potency-feeling, like the assent factor, is a kinesthetic sensation or memory'.

In the next stage of the exposition the dynamic potency attributed to incipient and remembered sensations of movement is rudely denied them: they have no potency to intensify thought or to make thinking hard; all potency is the potency of associative reproduction according to the principles of frequency and recency of repetition of a sequence of ideas or sensations impressed upon the organism by the chances of the environment. The origin and growth of every association 'may be attributed to chance concurrence, frequent repetition, and value'. The dragging in of the word 'value' here is as significant as it is absurd. Warren is engaged in an attempt to explain away all value: for value is a function of purpose; in a purely mechanistic world there can be no values. Yet at the crucial point in the mechanistic explanation of purposive action, Warren's own explanation rightly appears to him so thin and wofully inadequate that to the purely mechanical factors he adds 'value'; he might just as well have added, in place of value, 'purpose'. The explanation would have been no less adequate and more ingenuous.

The self-factor also 'may be traced to elements which arise during the process of fulfilment'; but 'the self-datum is the least important factor in the purposive experience. It is less characteristic than the assent, less vivid than the potency feeling'. It is allowed no potency, not even for the space of one paragraph. And 'though it should not be ignored' we need not trouble about it.

Finally, the fifth factor, the sense of fitness or unfitness. This 'feeling-element by itself does not seem to be in any way characteristic of the purposive consciousness; it is the same sort of experience as occurs in the satisfaction of any want or need'. And this 'sense of fitness' is not only a feeling, it is also a judgment, 'the judgment of correspondence between forethought and realization'. It is of no importance and therefore we need not insist that judgment is utterly out of place in a mechanistic association psychology, such as Warren expounds. Warren's analysis of purposive consciousness seems to me to require no further comment. I will only add that, if this is the sort of thing that results from the practice of introspective psychology, the exercises of the cruder behaviourists would almost seem preferable as an alternative. And I will point out that Warren, in reducing to kinesthetic sensations the various feelings and senses discovered by him in purposive consciousness, relies largely on the authority of the late Professor Münsterberg. It is therefore in order to remind the reader that, as I have set forth in detail elsewhere,[1] Münsterberg, having expounded through thirty years the sort of psychology accepted by Warren and having made himself the leading exponent of the reduction of all experience of activity to mere kinesthetic sensation, turned completely about and in his last book[2] frankly recognized that a psychology that is to be of any practical value, that is, a psychology that can answer to the pragmatic test, must be out-and-out purposive, rather than mechanistic. In all his earlier publications he had sought to purge psychology of anthromorphic (or, as Warren puts it, psycho-morphic)

[1] 'Purposive or Mechanical Psychology', *Psychological Review*, 1923.
[2] *Psychology, General and Applied.*

interpretations and to render it purely mechanical. His last book shows that he had realized the absurdity of excluding all trace of human nature from the scientific description of man; for that is what the mechanistic psychologies seek to accomplish.

After analysing ' purposive consciousness ', Warren turns to consider the objective marks of a purposive process and, neglecting all others, chooses as the essential ones, two, namely, anticipation or seeming preparation for later coming phases, and what he calls ' fitness '. He finds such anticipation to be very commonly displayed both in animal behaviour and in organic processes in general. He even goes so far as to say that, in such processes, the present course of events is conditioned by subsequent events, those which naturally follow. ' The actions of a dog performed while pursuing a rabbit are in a measure conditioned by the subsequent act of catching and eating his prey as well as by the antecedent stimuli—visual and olfactory.' And: ' Behaviour and growth are actually conditioned by the future as well as by the past. A later event is the basis of some earlier event, not merely its effect.' Such anticipation or preparation is, he says, ' found in all the higher types of behaviour, instinctive as well as intelligent. Even tropisms exhibit it. It appears in the phenomena of growth, both individual and phylogenetic '. What then does Warren mean by the expression ' in a measure conditioned by the subsequent act ' ? If we ask him: Do you mean ' caused by ' ? Do you mean that the subsequent event has causal efficacy upon the present course of events ? he would undoubtedly repudiate the suggestion. But it never occurs to him that perhaps the psychical or mental anticipation may have causal efficacy in present psycho-physical events, that it may be a condition of what he calls the objective anticipation or preparatory action. Instead he alleges that all who do not accept purely mechanistic explanations of human and animal actions (he speaks of them comprehensively as vitalists) regard ' the purposive relations as counteracting or superseding the causal relation. According to his [the vitalist's] view there are in purposive phenomena one or more " uncaused " terms.' That is to say he ignores the possibility that foresight of future possibilities may have causal efficacy, assuming that there is and can be only one kind of causation, namely, the mechanistic, that in which foresight plays no part. In other words the conclusion which he reaches by his elaborate investigation, namely, that there is no teleological causation, is contained in his premises by implicit definition, to the effect, namely, that all causation is mechanistic.

Warren then surveys the inorganic realm in search of any instance of anticipation and finds none outside the working of man-made machines. 'Anticipation is so marked a characteristic of biological phenomena, that the most mechanistically inclined biologist finds it difficult to dispense altogether with teleological language in discussing growth and behaviour. On the other hand, when we study the "behaviour" of chemical elements, the evolution of the stellar universe, and the formation of the earth's crust, it is difficult to discover a single indisputable instance of anticipatory activity.' Nevertheless he believes 'there is some ground for extending the concept of fitness into the inorganic realm', and that 'this extension, however, serves only to emphasize the mechanistic interpretation of purpose'.

Surely the fact acknowledged by Warren that anticipation, the conditioning in some sense of present events by future events, pervades all the organic world while no trace of it can be found in the inorganic, this fact should give him pause, suggesting irresistibly, as it does, the prevalence in organic nature of some form of causation absent from the physical realm! And, when we add to anticipation various other objective marks of purposive process and find that they have the same peculiar distribution in nature, and find also that, wherever introspective observation is available as evidence, we have evidence of foresight and impulse and satisfaction correlated with these objective marks, are we not justified in asserting that we have a strong *prima facie* case for the reality of teleological causation as a mode distinct from the mechanistic? For, though Warren arbitrarily asserts: 'Purpose in the last scientific analysis implies only anticipation and fitness', this assertion is plainly not true. It implies also the 'purposive consciousness' which he has analysed so unsatisfactorily; and it implies other objective marks not found in the physical realm; it implies persistence towards the goal with variation of means or route taken, a persistence that is independent of extrinsic influences and which ceases on the attainment of the goal.

I will point out one last inconsistency. After alleging that anticipation pervades the organic realm and is present in the phenomena of the growth and behaviour of lower animals, Warren asserts : ' Anticipatory behaviour depends upon a set of mechanisms, especially the distant-receptors and the central nervous system.' How then about the phenomena of growth and the behaviour of the lower animals that have neither distant-receptors nor central nervous systems ?

Professor R. B. Perry's Purposive Effort

Dr. R. B. Perry is a philosopher who has written extensively about purpose, and has written among other things a book on values.[1] He sees clearly enough that valuation is a function of our conative nature, of the desires springing from our instincts and sentiments ; that in a purely physical world there is no valuation and no value. He sees also that the evolution of civilization is essentially the production, conservation and accumulation of values, or as I would prefer to say, of valueds, i.e. of modes of action, of thinking and feeling, of forms of organization and of institution that have value for us ; that the higher efforts of mankind are efforts to conserve and increase the sum and the system of values, and that these efforts have been on the whole attended by a certain degree of success.

Yet in spite of all these just recognitions of the important facts of value, Perry will not admit the reality of teleological or purposive causation as a mode distinct in nature from the mechanistic, and in his book on values and in a series of articles does his best to exhibit purposive process as but a kind of mechanistic event.

I hardly know how to characterize this effort ; for I do not find any consistent scheme for mechanistic interpretation of purposive events. It seems to me rather to consist in an endeavour to smooth away the sharp edges of events of both kinds, the mechanistic and the purposive, until both, being reduced to vague and uncertain outlines, become indistinguishable.

In one article he is concerned to show that, when we execute a plan of action, we may validly assume that there is laid down in the nervous system a mechanism (or a co-ordinated series of mechanisms) which governs each

[1] *General Theory of Value* (New York, 1926).

step of action as the appropriate occasion arises. Now this is perfectly legitimate and plausible. Suppose I make a large bet that I will eat a particular fruit on each day of the week as the sun rises, an apple on Sunday, a pear on Monday, a peach on Tuesday and so on. It is a fair assumption that, when my plan of action is laid, there is formed in my brain a series of dispositions (call them neurograms or what you will) which may be conceived to work, according to the mechanistic laws of associative reproduction, to bring about the preconceived series of actions. But there are two fatal defects of the scheme. First, suppose that, waking a little before sunrise on Tuesday, I discover that some one has eaten my carefully preserved and only peach. I leap out of bed and ransack the pantry. I rush over to my neighbour to borrow one ; but he is away from home. I break open his garage, take out his car, and drive furiously to the greengrocer's store ; I hammer on his door and stave off with subtle words the policeman who seeks to obstruct me. Finally, after trying a dozen plans, I smash the window and seize my peach just as the sun is rising.

Here is a long train of varied actions all directed to the one end or goal ; and yet only the last of the series had a place in my plan ; all the rest are novelties. I have never before done such outrageous things, never conceived them. Yet such, though less extreme, variations from the preconceived plan of action are the rule, rather than the exception, in all typically purposive action.

Secondly, even if my plan works out perfectly according to schedule with a minimum amount of variation from and addition to the various steps as preconceived, there yet remain the most essential features of the purposive activity, namely, the volitional acceptance of the goal and the formation of the plan of action. Here foresight and conation do their peculiar and essential though imperfect work ; and it is here that the postulation of mechanistically formed and operating neurograms and associative neural links will not explain the facts, no matter how numerous they may be nor how complete their connexions with one another.

Although many, perhaps the majority, of modern psychologists have done their best, under the influence of the mechanistic prejudice, to ignore the purposive nature of human action and mental life, the facts are so insistent that some degree of recognition of them and some attempt to explain them can hardly be avoided. And we find accordingly that some recognition, however grudging and partial, is made by all but the extremists. And, since it is in bodily action that purposiveness appears most insistently, the recognition is often restricted to that sphere. Such psychologists have fallen into the way of postulating what they call a ' motor set ' or a ' determining tendency '. And this motor set is supposed to be something in the nervous system that is formed and operates mechanistically. No one attempts to describe either its nature or the process of its formation, or its mode of operation. It seems to be generally conceived as a sort of charging of certain motor neurones such that they are made ready to discharge along their efferent axones upon any slight stimulation. That seems to be as near as we can get to picturing the ' motor set '. Now this way of conceiving a ' motor set ' was reached chiefly by considering the processes of time-reactions. A subject accepts an instruction to make a certain movement (such as to depress a finger) on receiving a certain signal. The signal comes and, without the intervention of any discoverable thought-process, the movement follows at a minimum interval. In such a simple case the ' motor set ' seems perhaps adequate. But suppose that the signal for reaction is to be a blow aimed at some part of me undefined, and that my task is to deal with it as effectively as possible. I cannot foresee just from which direction or at what part the blow will be aimed. And, when a particular blow does come, there are perhaps half a dozen ways open to me for dealing with it, a retreat, a counter, a dodge to left or right, a guard. The most intelligent boxer is the one who on the spur of the moment chooses the best movement for dealing with that particular unforeseeable blow. Of what use is the postulation of a ' motor set ' in this case ? If I form a ' motor set ', it may determine a most inappropriate movement, and I shall be knocked out.

It is one of the essential marks of the purposive action that it adapts itself indefinitely to the particular unforeseeable circumstances and reaches its goal by means of any one of a multitude of possible movement-combinations ; and the goal, the attainment of which alone terminates the action-sequence, is not in most cases any movement, but something to be attained by means of any movements whatsoever that may serve the purpose. How then can the postulation of any particular 'motor set' help us? Such a set would in very many cases be highly prejudicial; it is perhaps characteristic of the slow-witted obstinate man or animal who persists in trying to force his way where no way is. The 'motor set' could account only for kicking against the pricks or for butting at a stone wall.

Now Perry makes use of this notion of a 'motor set'. His mechanistic apparatus for the explanation of purposive action seems to consist in a 'motor set' governed in detail by the schedule or plan registered as a series of linked neurograms. But as we have seen, neither of these two pieces of mechanism will serve the purpose; and a conjunction of the two must be equally ineffective.

In his book on values, Perry makes fuller recognition of the peculiarities of purposive action. He recognizes that human and animal action is a seeking of goals. And he recognizes that all such action implies some 'governing propensity', a propensity to seek the goal, whether that goal is attainable only by means of bodily movement or is a mental goal, such as a forgotten name. Perry correctly describes in some detail the role of such a governing propensity or conative disposition in action. He writes: 'Nature here rises to that level which we recognize as characteristic of mind or intelligence. . . . The crucial difference which marks this advance, *and which is the contribution of mind*, is control by anticipation; which constitutes the essential meaning of what is variously known as prescience, prospicience or foresight.' Again: 'However he may have come by it, the animal is supposed at the moment of action to possess a capacity for prospiciently determined action. He acts not because of what is or has been, merely, but because of what he anticipates. . . . The response anticipates its own proper object. . . . The dog who anticipates a beating is not merely behaving in a way that averts a beating, but is in some measure presently enacting the behaviour appropriate to the possible future beating which the whip represents to him. . . . The situation is construed by the agent in terms of something ulterior. . . . That which is lacking in the strictly biological picture has been provisionally termed "prospicience", and consists in the capacity to act in the light of expectation.' Again he writes: 'That aspect of life which is progressively characteristic of animal organisms and pre-eminently characteristic of man is not adaptation, but *adaptability*. It consists not in an equilibrium between the existing organization of the species and the constant features of the environment, but rather in a capacity to form projects, deal with novel situations, overcome difficulties, and *plan ahead*. . . . An organism in so far as acting interestedly is *always* docile and experimental; the controlling propensity is always capable of inciting to new and untried efforts and of exercising a selective function with reference to the tentative acts which it instigates. . . . That which is indispensable to intelligence is not that behaviour should reflect the past, but that it should anticipate the future. Its adaptive character will be in the agreement between this anticipation and objective causal connexions. . . . As in the case of animals, so also in the case of man, we shall discover the factor of interest or purpose in the *modus operandi* of these governing propensities; and, more specifically, in their future reference, and in their tentative selection of subordinate responses.'

All this is excellently said; the reference to the future course of events, which, as we have seen, is the essential mark and condition of teleological events, is fully recognized. But he assures us that 'As prospicient adaptation, interest may be regarded as "teleological" without implying any breach with "mechanism".' That is to say, he regards the teleological events which he so well describes as essentially mechanistic; but he does nothing to show how an action that is governed by prospective reference to events still in the future can be mechanistically explained. In a paragraph entitled, 'The Reconciliation of Mechanism and Teleology', he suggests that the acceptance of the principle of emergence [1] effects this reconciliation. Under such acceptance 'the mechanism of any system would mean simply its composition and structure. And, *per contra*, teleology would be taken to signify simply the novelty of the synthetic properties; or, the emergent character would be regarded as the purpose of the conditions from which it emerged. Whichever of these interpretations was accepted, mechanism and teleology would no longer be conflicting hypotheses.' This is all we are given to justify the denial of all radical difference between mechanistic and teleological events. Yet it is clear that neither of the two

[1] Cp. chapters V and VI.

alternative definitions of the teleological offered us in this passage takes into account the essence of teleological events, namely, their governance by prospective reference.

The process that, it would seem, suffices to justify to Perry the denial is the identification of the 'governing propensity' asita 'physiological disposition'. Now a conative dispos ion must unquestionably be regarded as physiological, in so far as it is a part of the organization of the organism, a part that plays a role in its functioning. The adjective 'physiological' is applicable to any process or structure in the organism that is concerned in the functioning of the organism. But Perry, like so many others, uses the term 'physiological' in a question begging manner; namely, he implies that a 'physiological' disposition is wholly and purely a material structure and must therefore operate according to the laws of mechanism. But that, of course, is exactly the question in dispute. The great question of the reality of teleological causation is not to be settled in summary fashion by this question-begging use of the word 'physiological'.

The Complacency of Dr. R. B. Raup

The thesis of an ambitious and well-written little book[1] by Dr. Raup is well described in the introduction to it by Dr. W. H. Kilpatrick in the following terms: 'It is offered as the foundation of human behaviour. But much more than this. As if to be the one foundation of human behaviour were not a sufficiently ambitious claim, this conception of complacency is offered as one which will at the same time not only place human behaviour in the same continuous series with all other biological behaviour but even with all physical behaviour as well.'

What then is meant by 'complacency', the key to the understanding of all events? The word is of course used in a widely extended sense. Dr. Kilpatrick goes on to explain that 'this conception of complacency' proves 'at bottom to be identical with the equilibrium tendency in physics'; and rightly says that this is both surprising and startling. 'Human behaviour', says Dr. Raup, 'has a central characteristic which we have chosen to call the complacency tendency, and any bit of behaviour can be accounted for and described as some phase or function of

[1] *Complacency, the Foundation of Human Behavior* (New York, 1925).

this tendency.' Again: 'The tendency of behaviour is always towards the resumption of a condition of complacency which shall adequately supplant the one disturbed.'

'Complacency' is, of course, a psychological term allied in meaning to the word 'satisfaction'. But in the hands of Raup, whose purpose it is to explain all purposive action mechanistically, the psychological meaning sinks into the background and the 'complacency tendency' of organisms is identified with the tendency of physical systems to return to equilibrium after disturbance. The working of the tendency in man is illustrated by the case of losing one's hat: a previous condition of equilibrium or complacency is disturbed by this event, and the complacency tendency at once begins to operate and continues until the obtaining of a new hat has restored complacency. In a similar way, when one's stomach empties itself, a condition of complacency is disturbed and the tendency is aroused and works until equilibrium is restored by a meal. Complacency is 'the condition-from-which and the condition-to-which behaviour is always in process of moving'. So far so good. We have, in a new terminology, the statement of a fact too often ignored, namely, the fact that in all bodily and mental activity we have to recognize the working of some impulse that is normally terminated only by the attainment of a result of some particular kind. This is the great fact of conation stated without regard to its introspective or immediately experienced aspect.

The general fact is so obtrusive that many of the most mechanically-minded psychologists have now added to the mechanism of associative reproduction (which to a few belated souls still seems all sufficing) the recognition of what they call 'drive' or 'drives'. The word has come rapidly into favour, because of its mechanical flavour and implications. Its use seems to incur no risk of the charge of appealing to 'mystic potencies'. It suggests a factory full of machines driven by power-belts. But for most of those who use the word 'drive' in this connexion, it has no other meaning; it means merely the mechanistic prejudices of the writer, and some recognition of the objectively observable facts of conation.

Raup's theory of complacency is distinctly superior to the 'drive'; for it recognizes, however inadequately, the fact of some particular condition that must be attained by the organism as the natural termination of each conative train of events; it recognizes the great fact of conative unity, the fact that our activity falls into natural units, each a sequence of indefinite length that expresses or is sustained by some dynamic condition or source of energy peculiar to to itself. And Raup's theory is that this dynamic ground of any train of activity is in every case a process that is wholly physical and mechanistic, one that can be explained without reference to the future, and one that is also the dynamic ground of all inorganic events; namely, the 'equilibrium tendency'.

What is this 'equilibrium tendency' in the physical world? It is said to be an expression of 'the second law of Energetics—that free energy tends to diminish itself, and bound energy tends to be at a maximum', or, alternatively: 'Where there is a difference of intensity of energy the tendency is for the higher intensity to diminish, and for the whole performance to result in a condition of balance, or equilibrium.' It is also expressed as the law that energy flows always from higher to lower potential, or as the 'Law of the degradation of Energy' or of entropy. Raup seems to state a well-established law of the physical realm when he writes: 'All change must in some way be traceable back to this basic law of the behaviour of energy.' But his theory makes two assumptions, both of which require careful examination: first, that all organic change must also 'be traceable back' to this same law of energetics; secondly, that in organic changes, taking place in conformity with this law, foresight of future possibilities makes no difference to the course of events, that the course of events is explicable without reference to the future. It may be that, even though the former be true, the second is false. In pursuing this inquiry we must not seek to take advantage of the fact that Raup, like almost all mechanists, in discussing physical processes, frequently uses anthropomorphic terms that in strictness are illegitimate, such terms as 'seeking', 'quest', 'optimum', terms which imply foresight and value and are meaningless when applied to non-purposive things and events. We merely note that the use of such words serves to mask to some extent the essential differences between organic and inorganic, between purposive and physical events.

Let us notice first that, according to the theory, the production of organisms, all the evolution and growth of their complex structures, including the brains of men, takes place as a series of events in the course of this one fundamental dynamic tendency, the 'quest' for equilibrium. Now the condition of complete and final equilibrium to which all events tend is (according to the theory) one in which energy is equally diffused at the same potential everywhere, in all systems and in all parts of them; one in which all chemical affinities are satisfied in the highest degree; one in which no stores of potential energy remain. But the whole of organic growth and evolution is a building up of highly unstable systems, systems characterized by great stores of potential energy held in reserve, stores which, as soon as life ceases in the organism, very rapidly follow the law of equilibrium or katergy, but which, so long as life continues, follow in the main an opposite law, the law of anergy; that is to say, so long as the organism lives, energy is constantly raised within it to higher potentials and stored as potential energy. Now it is true apparently that, as somewhat rare exceptions, anergic processes occur in the physical realm, as incidents of larger katergic events. But the fundamental fact of the processes of organic growth and evolution is that in them anergy predominates over katergy; the anabolic over the katabolic processes. And thus the organic realm considered at large presents itself as one vast exception to the law of katergy that rules the physical realm. Since, then, there is for every organism and for all its parts a short and easy way to equilibrium, the way followed as soon as life ceases, is it not *prima facie* a wildly improbable assumption that all life, all growth, and the whole process of organic evolution through the ages is but the expression of the equilibrium tendency. What a round-about way to the equilibrium of death, to evolve, during some hundreds of millions of years, organisms of increasing complexity, the total effect of whose existence is to postpone by that period the attainment of equilibrium!

If we regard broadly the life of the individual organism we discern a similar incompatibility with the theory. The theory asserts that all things seek equilibrium, and all vital activities are merely incidents in this 'quest', in this trend towards equilibrium. Yet the nature of all vital or organic processes is best summed up in one famous phrase, the struggle for survival, i.e., the struggle against death, the struggle to stave off the attainment of equilibrium.

Again, the major part of the activities of any animal fall under two heads, the quest for food and the effort to reproduce and perpetuate its kind. Each of these is a distinctly anergic process; that is to say, in so far as successful, the process involves the predominance of anergy over katergy; for the food-quest results in the building up of new and greater stores of potential energy; the reproductive activities result in production of new highly unstable systems rich in potential energy and resistive to the equilibrium tendency.

May we not rather say, then, that Life, whatever else it may be, is something that works against the equilibrium tendency, some organization which, though it is in effective relatedness with the physical energies that perpetually flow through it, 'seeking' equilibrium according to their fundamental law, works to check this tendency and postpone its realization? Do not the empirical facts point to the view that living organisms are the seat of events that obey some other law than that of equilibrium, one antagonistic to it?

We see the same conclusion indicated even more clearly by the facts of human life at its highest. We see an Alexander setting out to conquer the world; a Roosevelt dedicating himself to a life of political strife; an Amundsen resolving to see both poles and to sail the North-west Passage. We are told that Amundsen formed these resolutions at the age of fifteen years. He studied the exploits of Sir John Franklin, and, says he, 'A strange ambition burned within me to endure those same sufferings. . . . I irretrievably decided to be an Arctic explorer. . . . My career has been a steady progress toward a definite goal since I was fifteen years old. Whatever I have accomplished in exploration has been the result of lifelong planning, painstaking preparation and the hardest kind of conscientious work.' Such is Life in its highest expressions, those in which its nature is most clearly revealed. Is Amundsen's career explicable as an expression of the equilibrium tendency through channels of least resistance? Was it not rather a perpetual and triumphant struggle against the equilibrium tendency of the physical realm, until in the end, his powers diminished by age, it became too strong for him? Does Mussolini's career look like a seeking of equilibrium, a mere dissipation and degradation of energy? And we may ask the same question of every instance of synthetic creative human activity. 'The nervous system is just a more quickly and more sensitively conducting part of the organism serving it in its adjustment with the surroundings in such a way as to facilitate the reduction of this condition of strain.' If this is true, can the mechanics of the brain explain the actions of the man who sets out to reach the pole, to march on foot across darkest Africa, or to scale Mount Everest? Is it credible of even the most ordinary mountain-climber that his foresight of standing on the summit has nothing to do with his arduous ascent, has no causal efficacy; that he is merely the creature of a purely mechanistic tendency for energy to degrade and dissipate itself? Surely experience proves conclusively that there are easier ways of dissipating human energy.

Raup's answer to all such questions as I have here raised runs as follows: All instances of long-range human achievement are instances of the fulfilment or striving for fulfilment of desire of one kind or another. Now desire is the expression in consciousness of a disturbance of autonomic equilibrium; and all the activities sustained by any desire (as we 'anthropomorphically' say) are in reality expressions of the working of the equilibrium tendency towards the restoration of that disturbed equilibrium. 'The status of affective rest is the end state of the dynamic striving . . . the process of neutralization of the affective disturbance is the dynamic principle underlying all behaviour. The behaviour observed at any period is the resultant or compromise of the various affective trends active at the moment and is always symptomatic of the affective state.' In the case of Amundsen, the reading of Sir John Franklin's exploits disturbed a condition of complacency, i.e. of equilibrium in one or more of Amundsen's autonomic segments; this equilibrium could be restored only in one way, namely, by visiting the South Pole and the North Pole and by sailing the North-west Passage: hence Amundsen's career. And if we still persist in asking—in face of this triumphant demonstration: How then explain Amundsen's final visit to the polar regions? the answer is easy. We have it on the high authority of Professor John Dewey that all human activities are habits. Going to the polar regions had become a habit with Amundsen. He simply couldn't help himself. He died a victim of habit.

Now, when we reach this stage of Dr. Raup's discussion, the stage where he begins to come to close quarters with human action, we find that the physical principle he invokes as all sufficient is no longer the second law of thermodynamics, the law of degradation of energy. He is now using the term 'equilibrium tendency' in a new sense, namely, the sense of the tendency to self-restoration of a system in dynamic equilibrium.

That living organisms are systems in dynamic equilibrium seems to be true. They have the power or tendency to self-restoration after disturbance by forces from without. But it seems certain that this 'equilibrium tendency' is something other or more than the basic equilibrium tendency with which our author set out. The instances of this kind in the physical realm are few and far between. Raup cites whirlpools, candle-flames, and waterfalls as examples. I doubt whether these are good instances. But a clear instance is, I suppose, a spinning top or a gyroscope in action. Push it from its plane or axis of rotation and it resists your push; and, when the push ceases, it returns to that plane or axis. Now, if it be true that this strictly mechanical principle is the key to all organic and mental phenomena, we are in face of a very remarkable fact, one which cannot be too clearly asserted. It is remarkable among other things in that it directly contradicts all the many statements we have lately had from high authorities who, in expounding *Emergent Evolution*,[1] agree in insisting that not only vital events but chemical events also are not explicable in terms of purely mechanical principles.

The question arises whether, outside the organic realm, any system of chemical dynamic equilibrium is known; for that is what a living organism is. Is there any inorganic system in which disturbances of chemical equilibrium are regularly followed by restoration to the *status quo ante*?

Raup adduces no such instances. I can think of none. I imagine that none is known. Dr. R. S. Lillie, on whom Raup largely relies, writes: 'This general conception of living matter, as a system which holds its own through a balance of constructive and disintegrative processes, is fundamental in physiology.' And also he writes: 'Under the terms regulation and adaptation, we include in their broadest application, all those features of adjustment—structural, chemical and active—which are especially characteristic of living as distinguished from non-living

[1] Cp. chapter V.

systems.' From which it would appear that in Lillie's view instances of dynamic chemical equilibrium are not to be found outside of living organisms.[1]

Raup writes of cell-metabolism: 'The cells are worn down, but there is also a corresponding heightened process of building up the cell. This metabolic process goes on just because there is activity in the part, throwing the energy condition off of equilibrium, and thus compelling a drive back toward that state.' Now this, of course, begs the whole question at issue, namely: Can this automatic restoration of chemical equilibrium be explained on physical or chemical principles alone? To say that katabolism compels an anabolic drive back towards the disturbed state may be true; but to offer it as a physico-chemical explanation (unless we can point to some analogous self-compensating chemical mechanism in the inorganic sphere) is absurd. Raup indicates no such analogy; and, as we have seen, Lillie, a high authority, implies that there is none.

Instead Raup flies off to another type of explanation of human action, what I would call the vulgar error of the 'action-pattern'. An action-pattern is conceived as a nervous mechanism such that, given the incidence of an effective stimulus on its receptor terminals, a certain movement must result. The simple animal is supposed to have few and simple action-patterns. Natural selection secures the increase of action-patterns in complexity and number, until the brain of man contains a very large number; hence the variety of his responses to stimuli. It is a widely accepted theory of human action; one that is readily accepted by simple minds. It does conform to the law of degradation of energy; but has nothing to do with, is not in any sense an expression of, the tendency of organisms to anabolize stores of chemical energy after katabolic expenditure. Given the action-pattern, it will produce its specific movement-combination whether or no anabolism subsequently restores the chemical condition. It is only subsequent repetition of similar action that depends on the restoration. But the action-pattern theory, in spite of its

[1] It is highly significant of the present trend in biology that this high authority, who formerly wrote as a strict mechanist, has recently published a series of articles in which he repudiates the adequacy of mechanistic principles to human life, and even finds room for indeterminism in the inorganic realm. (Cp. especially 'The Scientific View of Life,' *Journal of Philosophy*, 1928.)

popularity, is utterly inadequate. What action-patterns will account for the career of Amundsen? What action-patterns will account for the long series of movements, all adapted to special conjunctions of circumstances, by means of which a dog finds his way home from a distance, drives a flock of sheep through a gateway, or opens a puzzle-box to find his biscuit inside it; or for the series of movements by means of which an ape, for the first time, joins two sticks together in order to rake in a banana placed beyond the reach of either stick? The action-pattern theory satisfies only those minds that have not grasped the nature of the problem to be solved.

Raup, of course, recognizes that many new action-patterns are acquired or built up as habits in the course of life of any intelligent animal. And he accepts the common view of the mechanical formation of habits which is so largely based on Professor Thorndike's famous experiments with hungry cats escaping from cages to find food. Now it cannot be too often or positively repeated that Thorndike's reports and conclusions are gravely misleading. All steps in animal psychology, all series of significant experiments on animals made since the date of that report, converge to that conclusion. I would refer more especially to the work of Dr. Yerkes on apes, of the *Gestalt* psychologists, especially Professor Köhler, on apes and other animals, to various articles by Professor E. C. Tolman, and to my own observations on dogs and rats. All these and many others show conclusively that the animals do not solve the problems set them by repeating successful movement-combinations that in the first place are strictly random. They show, first, that the animal is striving towards a goal, is directing its movements towards the goal; secondly, that there comes very often a moment of insight, a moment when the animals grasp the relations involved in successful movement and that, from that time on, they have an understanding of such relations which enables them to deal effectively with different but similar complexes of relations. And this insight or understanding cannot be interpreted in terms of any action-patterns, inborn or acquired.

Dr. Raup's effort, then, brings together a number of very different principles of explanation, some of which are irrelevant, others false. The one that has most claim upon our attention as a general principle explanatory of human and animal action is that which postulates in organisms a tendency to restore their condition of chemical equilibrium after disturbance and regards all actions as incidents in such restoration. Now this theory has been elaborated by Dr. E. Rignano in a number of publications of earlier dates than Raup's book. He may properly be regarded as its sponsor and creator.

I shall therefore examine it more nearly in the form in which it is presented by Rignano.

I will add only one further general criticism. When Raup comes to deal with the higher forms of human activity, he is led by his several distinct principles of explanation to make the following generalization: 'Man values positively (favours or chooses) that which arouses disturbances for which he has reduction-patterns ready. He values negatively (holds in disfavour, rejects) that which arouses disturbances for which no reduction-patterns are ready in time, or readily available.'

In the term 'reduction-patterns' we see an attempt to synthesize the principle of the action-pattern with that of the reduction of disturbance of equilibrium. The generalization is deduced from the several theories; and, since the generalization is manifestly false, it confirms the view that the theories which have led to it are false also. I say the generalization is untrue. It amounts to saying that in general man prefers and chooses those things and situations that require of him only habitual reactions and turns away from whatever is novel and requires novel modes of action. This may be true of a limited number of elderly persons made cautious and conservative by much experience of trouble and disaster. But we know how all others quickly grow dissatisfied with the familiar and the habitual; how novelty of any sort appeals to them; how they seek after new places, new faces, new fashions, new practices, new theories, new gods, are lured on by every promise of novelty, of change, of mystery, of wonder and awe. Why do a million Americans swarm into Europe every summer? Why, rather, is the swarm not one hundred million strong? Only because they have not yet sufficient money for the journey. Man's most striking characteristic, perhaps, is his insatiable appetite for the new, the strange, the thrilling. It is true that he needs a background of the familiar and habitual to which he may return for rest and recuperation, where he may by predominance of anabolism restore the energies depleted by excess of katabolism. Even Amundsen needed head-quarters and a home.

Dr. Rignano's Theory of the Affective Tendencies

Dr. Rignano's theory of the affective tendencies has been elaborated in a number of publications extending over many years and restated in his *Psychology of Reasoning.*[1] He deserves much credit for his insistence on the importance of what he calls 'affective tendencies' and others of us prefer to call 'conative tendencies or impulses'. He rightly insists that such tendencies are involved in and are the dynamic sources of all human activities.

His theory seeks to account for the phylogenesis and the operation of these tendencies so generally ignored by the mechanists. He rightly sees that this is the first and essential problem for every mechanistic theory of human action. In his recent book he aims to show 'the close connexion which unites this particular manifestation of vital teleology, reasoning, with the very essence of all biological finalism in general'. But his theory of the affective tendencies is an attempt to find a mechanistic explanation of teleological or purposive activity in general.

'Observation of the behaviour of the various organisms from the unicellular up to man, shows that a large number of their movements or acts, and especially the most important ones, may be interpreted as manifestations of a tendency of the organism to maintain or to restore its "stationary" physiological state (to use Ostwald's terminology).' And Rignano's effort is to show that the working of every affective tendency may be explained as an instance of this general physiological principle. Now this tendency 'to maintain or to restore its stationary physiological state' is identical with what Raup calls 'the tendency to restore dynamic equilibrium'. We have seen that there are two questions to be satisfactorily answered before we can accept this tendency as the mechanistic explanation of all purposive events: firstly—Is such a tendency manifested in the inorganic world? Secondly—Does it account for purposive action in men and animals?

We have seen that Dr. R. S. Lillie regards the tendency of the times to restore their chemical equilibrium as one peculiar to living organisms. Rignano takes the same

[1] London, 1923.

view. He regards it as a function which is in turn due to a fundamental peculiarity of organisms, a peculiar property in which their nervous energy plays the essential part and for which he proposes the name, 'mnemonic property'.

This fundamental mnemonic property consists in this: any tissue, subjected to certain conditions which force upon it a certain type of metabolic change, acquires after a time the tendency to continue to metabolize in the same way; this is adaptation to the conditions imposed on it. Every organism thus becomes a system of such tendencies to maintain and restore the form of metabolism primarily imposed upon it by long-continued exposure to certain environmental conditions. 'Every organism is a physiological system in a stationary condition and tends to preserve this condition or to restore it as soon as it is disturbed by any change occurring within or outside the organism. This property constitutes the foundation and essence of all "needs", of all "desires", of all the most important organic "appetites". All movements of approach or withdrawal, of attack or flight, of seizing or rejecting which animals make are only so many direct or indirect consequences of this very general tendency of every physiological condition to remain constant. We shall soon see that this tendency in its turn is only the direct result of the fundamental mnemonic characteristic of all living matter.'

Thus hunger or food-seeking is the working of the tendency to restore the normal condition of the blood and other tissues depleted of chemical reserves. Sex activity is the working of the tendency to get rid of accumulated secretions of the sex glands. Maternal behaviour expresses the need to get rid of the secretion of the mammary glands. Rignano cites a number of relatively simple instances of adaptation, and writes: 'They show that the new physiological state arising from adaptation to the new environment, when once it has supervened and has lasted for a certain time in the organism, tends to renew itself. This tendency of a past physiological state to "re-activation" or reproduction is merely the tendency inherent in every mnemonic accumulation to "evoke" itself again. Hence it is a tendency of a purely mnemonic nature. But from this it would follow that the tendency to physiological invariability from which originate, as we have seen, the most important affective tendencies of all organisms whatever must be equally mnemonic in nature.'

More specifically, an ' affective tendency ' is a mnemonic accumulation of energy ; and it consists in an elementary specific accumulation of matter so constituted that it tends to restore itself whenever disturbed. Now, though Rignano expounds his theory in physico-chemical terms in the main, there is the one term of ambiguous import which occupies the central position, the term *mnemonic*. We seek in vain to discover from his pages whether this is a physico-chemical or a mental function or property. But at least it is clear that he regards it as peculiar to living organisms ; and that, for him, it is the ground of all the essential differences between organic and inorganic events.

'This mnemonic property, this faculty of "specific accumulation", whose absence leaves inorganic nature exclusively in the power of forces *a tergo* and deprives it of every finalistic aspect, is on the other hand everywhere present in organic nature and because of its presence makes the world of life a world apart, a world whose most essential features, the physico-chemical laws alone, in the limited sense assigned to them to-day, are quite incapable of explaining.'

It remains, then, doubtful whether Rignano's theory should be classed as an attempt to find a mechanistic explanation of teleological events. The language he uses leaves us in doubt. It is true that he uses many words that imply the causal efficacy of foresight and striving ; but most of the avowed mechanists do the same, while repudiating the teleological implications of such words. For example, he writes : ' Essentially then the " will " is nothing else than a true and proper affective tendency which inhibits other affective tendencies because it is more far-sighted, and which in its turn impels to action like all affective tendencies.' And he cites with approval the following words of the late Dr. Maudsley, one of the most outspoken of the old-fashioned materialists : ' There is present in the action of will some desire of a good to be obtained or of an evil to be shunned, which imparts its driving force.' He writes also that affective tendencies must be regarded as ' tendencies or aspirations of our mind towards a certain end to be attained '. And ' this affective activity, thus seen to be the great artisan of our intelligence, the influence which spurs it on and restrains it at the same time, is itself due to the fundamental mnemonic property, is, moreover, the most direct and characteristic manifestation of this mnemonic property of living substance. So that the mnemonic faculty, which, as we have shown elsewhere, explains the most fundamental biological phenomena—from the predetermined morphological adaptation of animal organisms and their behaviour—instinct with its unconscious foresight, to the transmissibility of acquired characters with its direct consequence in phylogenetic evolution and ontogenetic development—this mnemonic faculty is now revealed as able of itself to furnish us with all the most varied manifestations of the psyche.'

Rignano nowhere expresses his view on the psychophysical problem, nowhere tells us whether he believes foresight to be effective in guiding the working of the affective tendencies to their goals. Yet his language implies that he would deny causal efficacy to the psychic accompaniments of the brain-processes, and justifies the statement that the 'mnemonic property', although it is postulated as the unique property of nervous energy, it is yet in all its nature and workings a physical and mechanistic principle.

'We have already seen that this mnemonic faculty may be defined as the capacity of reproducing, through internal causes, the specific physiological states, which primarily required for their production the action of the forces of the external world. We have also tried to give an accurate account of the mechanisms involved, by admitting nervous energy at the base of every vital phenomenon and endowing it with the property of specific accumulation, that is to say, by supposing that each nervous accumulation gives as a "discharge" only the same specificity of the nervous current of "charge" by which the accumulation was stored. . . . In order to account for the most fundamental biological and psychological manifestations of life it is sufficient to suppose that there exists in nervous energy, over and beyond the properties common to all the energies of the inorganic world, nothing more than the mnemonic property. . . . This mnemonic property is exactly what gives to life its finalistic aspect, which consists in being moved by forces *a fronte* rather than by simple forces *a tergo*. The end towards which man gravitates with his affective tendencies, the future external circumstances which the animal unconsciously prepares itself to face with the complex behaviour that instinct dictates, such and such a peculiarity of environment relatively to which the organ fashioned by the embryo in the maternal uterus will be adapted, now function as a *vis a fronte*; and they function thus because they were in the past *vis a tergo*, and because the physiological activities, then determined by them in the organism, have left of themselves a mnemonic accumulation which now constitutes in turn the true and actual *vis a tergo*, directing and moving the development, the instincts and the conscious conduct of the living being.'

I cite these many passages because they show so clearly this distinguished thinker labouring earnestly and honestly with the problem of teleology. They show that, in the end, his explanation of purposive action is mechanistic. The explanation offered is in terms of a postulated peculiarity of nervous energy; and in the end the working of this energy is not guided by foresight and insight, but is a *vis a tergo* that directs the conscious conduct of man. The mental remains a mere shadow, a passive epiphenomenon of the mechanistically working physico-chemical brain-events.

Our two questions then are in order. First, is the postulated tendency or function one that we can validly conceive as a physico-chemical process? It is admitted, it is asserted by Rignano, that it has no parallel, no analogue, in the physical realm. Rignano does not use the expression 'emergent evolution'; but we may ask: Is nervous energy, manifesting this peculiar mnemonic property, an emergent from the physical realm? That seems to be implied by Rignano; though he does not discuss the question of its relation to other forms of energy. If the mnemonic property is an emergent, it is not a property of some primary or fundamental form of energy. If it is not an emergent, it must have entered our world when Life began and have continued to flow in in increasing quantities as life multiplied on the face of the earth. We cannot positively rule out either view; but both seem wildly improbable; and, until Rignano shall declare himself for one or other of these alternatives, it would seem unprofitable to discuss them.[1]

We raise our second question. Granted the existence in organisms of 'specific mnemonic accumulations' laid down through this highly peculiar operation of nervous

[1] Cp. chapters V and VI on 'Emergent Evolution'.

energy, would they, together with the mechanisms of associative reproduction, render possible mechanistic explanation of purposive intelligent action ? Here again Rignano has the great merit that, unlike most of the mechanists, he does not ignore or deny the more important of the facts to be explained. He recognizes the great peculiarity of purposive action, namely, its infinite adaptability to particular circumstances ; the fact that purposive striving, nonplussed in one direction, seeks another and another line until it attains its goal. To the affective tendencies he assigns a peculiar ' fundamental character ', namely, ' that they constitute a force determining, so to speak, a definite end to be attained but leaving undetermined the path to be followed '. I draw attention to the significant words ' so to speak '. Do they not imply in the author's mind a doubt of the validity of his assumption ? ' It is only when one series of movements happens to bring the organism back to the desired environmental relations earlier than another, that it will *from that moment* be " preferred " above the others ; and this may be expressed by saying that the affective tendency has exercised a " choice ". Hence it is only from that moment that the affective tendency will by mnemonic association constitute a force which " impels " these movements towards the end. . . . Until this takes place, however, the affectivity has no tendency at all to discharge in one path rather than another. Hence the great difference between the affective tendency or act of will on the one hand, and the reflex movement on the other. The reflex movement . . . represents, that is to say, a tendency to discharge along one single path determined in advance. . . . On the other hand, the affective tendency constitutes a force of which neither the point of application nor the direction are predetermined, but only the point towards which it tends. It is a ' disposable ' energy to be applied indifferently to this or that act so long as it leads to the desired end. . . . The reflex movement admits therefore of only one solution. On the other hand, the affective tendency admits of *an indefinitely large number of solutions*. . . . This possibility of many solutions constitutes the "unforeseen ", or " anti-mechanical " aspect in affective or volitional behaviour.'

Exactly! But, given an affective tendency, the tendency of a specific accumulation of potential energy to renew or restore itself, is it consistent with physico-chemical or mechanistic principles to assume that it may determine equally well any one of an indefinite number of different forms of bodily and mental activity, and may determine a series of these in succession, all directed to the goal, and to continue until the goal is reached and the chemical accumulation thereby restored ? Is there in the physical realm any analogue to such a series of events ? Water running down the face of a hill may seem to present some analogy. We may say that its goal is the sea-level and that it seeks and uses every available channel in turn until its goal is reached. And we may improve the parallel by pointing out that if on successive occasions water be poured on the same starting-point, it will tend to follow the complex channel ' chosen ' on the first occasion ; that with repetition the channel becomes more fixed, though liable to improvement by deepening and shortening under repetition of the flow of water.

This, I suggest, is the nearest we can come to finding a physical analogue for the infinitely variable and self-adapting course of purposive behaviour. But is the flow of water truly analogous ? I think not. No matter how many turns and twists the stream of water may make in reaching sea-level, it obeys at every moment one simple law, that of gravitation. Rignano and others who do not ignore the goal-seeking nature of human and animal action frequently speak of the animal as ' gravitating towards its goal ', and imply thereby that they would regard the flow of the water as truly analogous. But compare the flow of the water with the behaviour of man or dog returning home from a distance through unfamiliar country. Like the water, his direction is liable to be changed many times by obstacles encountered. But, unlike the water, his course is not explicable by any physical law. Finding himself in a cul-de-sac he may retreat, travelling away from his goal; he may return to his starting-point and renew his effort in a new direction. And always his desire for or urge towards his goal is rendered effective only by his knowledge of the direction in which it lies, by his insight into the spatial relations and by his foresight of his own actions. Neither a mere blind tendency to move, a mere restlessness, nor a steady pull or push in any direction, undirected by insight and foresight, will bring him to his goal. It is true that the purposive action implies within the organism a liberation of energy, a transformation from some potential to an active form, a liberation that continues until it is brought to an end by the attainment of a particular change of the internal condition. And for such liberation Rignano's postulated 'specific accumulation' provides. But this alone would account only for ei a specifically determined movement series (such as theory of the action-pattern provides for) or for some utterly blind and random movements. Purposive action, on the other hand, however lowly, is directed towards a goal however vaguely, even when it is a mere groping, as of a man groping in the dark of an unfamiliar chamber. He gropes, we may say, blindly; yet not quite blindly, not unintelligently, not in purely random fashion: for he is groping for an exit, even though there may be none; and his groping is directed to some degree by insight and foresight. Without effective guidance of this sort there is no teleological causation. For such guidance Rignano's scheme, occupying as it does an ambiguous position between mechanism and teleology, makes no provision.

I will add only that in the year 1926 Dr. Rignano made an address before the British Association for the Advancement of Science. In that address (which I heard but have not yet been able to read) he proposed to explain purposive action by postulating in organisms a form of energy that works teleologically. This proposal seemed to me to be an admission on his part that the scheme previously formulated is inadequate to his purpose.

Purpose and the Psychologists of the 'Gestalt' School

It is the aim of the psychologists of the school of *Gestalt* or configuration [1] to reform psychology by superseding the inadequate mechanical principles which have so long been in favour as explanations of animal and human behaviour and of mental life, the principles of the compound reflex, the conditioned reflex, the action-pattern, the principle of associative reproduction and of atomic sensationism. They have been mainly concerned hitherto with the facts of perception, though not exclusively. They have brilliantly and conclusively demonstrated the inadequacy of the principles they would supersede. The

[1] I write here of those of the Berlin group. If I understand rightly, those of the Graz group take a position more like my own.

principle they would put in the place of these is the principle of *Gestalt* or configuration.

A configuration is a system of energies in which every part co-operates in determining the whole, and the whole in determining every part. The whole system so considered is other than and more than a mere resultant or sum of its parts; it is, as another school would say, an emergent exhibiting properties that are not to be found in any of the parts.

Any field of perception, they say, is such a configuration; it is not merely so many distinguishable sensation-qualities juxtaposed in time or space; every part is what it is only in and as a working part of the total configuration. Thus within a heard melody each tone plays its part in constituting the whole; and it takes from the whole something of its peculiar quality as a tone in that melody, something that makes it other than the same note heard in isolation.

So far, so good. Now the *Gestalt* psychologists, and especially Professor W. Köhler, are equally concerned to show that the principle of configuration holds good of physical systems of energy, say a magnetic field, or a system of surface-tensions such as a soap bubble.

When they turn to consider human and animal action they recognize that it objectively exhibits tendencies towards goals. Their recognition of the fact is somewhat grudging and inadequate. Still it is much that it is there. They recognize the fact by saying that each configuration has a tendency to 'closure'. This seems to mean that any such system has a natural distribution of its component energies which is one of equilibrium, and that, when it has been disturbed by the incidence of extrinsic forces, it tends to resume that distribution or position of equilibrium. They say, then, that the tendency of any action towards a goal is an expression of this tendency to closure, which prevails equally in the physical and the mental realm. In this way they make it seem that purposive action is not radically different from certain purely physical processes, although they deny that such processes are explicable in strictly mechanical terms. In other words, their explanation of purposive action is not mechanical, but it is mechanistic in the sense we have defined; and it enables them to abide by the principle of Psycho-physical Parallelism, if they so wish.

Also they have recognized the fact and the importance of *insight* in all learning processes, seeing clearly that the mere formation of habit by repetition of random movements is something entirely different from intelligent learning. But they have not yet recognized the equal importance of foresight; and, until they do this, their scheme will remain subject to the same criticism and verdict which I made on Rignano's scheme. Like that scheme, it remains an ambiguous compromise between mechanism and teleology; and, in spite of the compromise, it will not work. In their scheme 'insight' and 'closure' remain ambiguous terms awaiting further elucidation; and I venture to prophesy that such elucidation will reveal the necessity of a frank recognition of teleological causation.

One member of this school, Dr. Kurt Levin, has published some very interesting experimental studies of impulsive and voluntary actions, with some theoretical discussions.[1] He arrives at conclusions closely allied to my own accounts of such actions. He recognizes the utter inadequacy of the explanations of human action offered by the mechanical association psychology. He rightly insists that, in order to understand any action, we have to seek in every case the source of the psychical energy involved. 'The experimental investigation of habit (association) has shown that the connexions established through use can never be regarded as themselves the motive power of a mental event; such a conception is erroneous also if one sees the essence of processes of habit and practice not in the formation of so many separate associations, but in the further development of particular action-systems. Rather in every case certain psychical energies (which as a rule can be traced back to some conative tension or need), hence certain psychical systems under tension (*Gespannte seelische Systeme*), are the necessary presuppositions of the occurrence of the psychical event, no matter what course it may take.'[2] Further: 'One has, therefore, to ask of every mental event, whence comes the causally effective energies.' He adds that 'the question whether ultimately we are concerned here with physical forces and energies can be left quite open'. Nevertheless, 'one must always

[1] Especially *Vorsatz Wille und Bedürfuis* (Berlin, 1926).

[2] These 'Psychical Systems' seem to be the instinctive or inborn conative dispositions which I have postulated as the energy sources of all mental events.

bear in mind that we are dealing with forces in the psychical field rather than in the physical conditions'. He insists that steering or guiding processes (*Steuerungs-vorgänge*) are of fundamental importance. Certain objects, he says, have a demanding quality or force (*Aufforderungscharakter*) which we experience as a lure or attraction (*Lockung*). Such qualities work upon us in the sense 'that they exert upon our mental processes, and especially our movements, a steering influence'. And such activities lead to satisfaction upon the accomplishment of the design and to the reduction of the tension of the underlying system to one of equilibrium at a lower level of tension. He speaks also of *hindrängen*, an untranslatable verb which expresses very well the working within us of desire; and also of active seeking for opportunities. Levin, then, comes very near to the recognition of the purposive nature of human action as something distinct in kind from all physical and mechanistic events. Yet he never crosses the Rubicon. Like the other members of his school, he shrinks from the decisive recognition of teleological causation; he does not even clearly formulate the question so urgently raised by his discussion, the question of the radical distinction between teleological and mechanistic causation.

The cautious non-committal attitude of this school towards our problem is well illustrated by a recent article by Professor W. Köhler.[1] He shows that certain physical systems are self-regulatory, i.e. each tends to a certain result in such a way that the result may issue in spite of disturbing influences from outside the system; and wisely adds: 'It does not follow from this that organic regulations rest on the same principles. For organic regulations have other peculiarities which have not been considered in our discussion and on account of which we shall perhaps be compelled to seek other principles than those of physical science.'

Other Contemporary German Schools

It is relevant to the discussion to point out that in Germany the *Gestalt* school is not the only one to revolt against the sterile mechanistic psychology that has so long dominated in academic circles.

[1] 'Zum Problem der Regulation', *Arch. f. Entwickelungsmechanik*, Bd. 112, 1927.

The psycho-analytic schools are exerting an increasing influence. Of these the Freudian school, although it has much to say of 'mental mechanisms' and professes 'strict determinism', is thoroughly hormic—that is to say, it recognizes that all mental activity, conscious or unconscious, is sustained by impulsions towards goals. The school of Adler is, if I have at all grasped its teaching, thoroughly purposive. The school of Jung is more frankly purposive and anti-mechanistic than the others. Many years ago (in 1912, I believe) Dr. Jung, in an address to the now defunct Psycho-medical Society of London, urged that psychology cannot hope to be scientific in the same sense as the physical sciences; that it must aim at understanding rather than at explanation. This was, I take it, one way of asserting that psychology must seek purposive or teleological interpretations, rather than mechanistic explanations.

In this respect Jung's precept is now followed by a school of German psychologists who expound *die verstehunde Psychologie* in which they are much concerned with motives, with motivation and intention. I mention, as representatives, L. Binswanger, Th. Erismann, G. Ewald, and Ed. Spranger.[1] The last-named especially, who seems to be the leading exponent of what is called *Geistes-wissenschaftliche Psychologie*, is exerting a wide influence through his well-known *Lebensformen* and other works, all of which repudiate mechanical explanations, and maintain a common-sense point of view.

Other influential contemporary writers in the German language (some of whom hold important academic positions, though for the most part they stand somewhat apart from the universities) who repudiate more or less explicitly the mechanistic psychology are Ludwig Klages, P. Häberlin, Max Scheler, Pfänder, K. Oesterreich, Hans Driesch and Hans Prinzhorn. The last-named has recently issued a number of publications (see especially *Leib-Seele Einheit*) in which he seeks to revive and carry further the thoroughly hormic psychology of Nietzsche.

In England, it may be added, the younger psychologists seem in the main very favourably disposed to the hormic view so ably expounded by Dr. T. P. Nunn. It would seem, in fact, that the mechanistic psychologists of America are

[1] Cp. Discussion on 'Understanding and Explaining in Psychology' by these four authors. *Proc. of Eighth International Congress of Psychology*, 1926.

in danger of finding themselves as isolated in science as their country is isolated in politics. It may be hoped that the warm welcome accorded to Emergent Evolution by a number of leading American biologists [1] may save them from this fate. Emergent Evolution with its ambiguous but anti-mechanical teachings is well suited to serve as a wedge that may pry open the door and let in a little diffused light without shocking their dark-adapted retinae. Let them take courage ; there is little reason to fear that their orbits shall contain ' Eyes grown dim in gazing on the pilot stars '.

[1] Cp. chapters V and VI and accompanying notes.

NOTE 7. *Dr. Rignano's Theory of Memory*

Considering ontogenesis, memory, and vital phenomena in general, Rignano writes : ' There may yet be the possibility that the resemblance which appears to exist between some of the essential properties of these three phenomena may be explained by a still more simple phenomenon which would form the common basis of all three catagories of phenomena ; the ontogenetic, the mnemonic properly so-called or psycho-mnemonic, and the vital. We have already shown that this common phenomenon might perhaps be implied in the hypothesis set forth above, in which we regarded the specific potential elements as accumulators of specific nervous energy or as specific elementary accumulators. In just this capacity of restoring again the same specificity of nervous current as that by which each element had been deposited one would look for the cause of the mnemonic faculty in the widest sense, which all living matter possesses. And further the very essence of the mnemonic faculty would consist entirely in this restitution.'[1] This passage states concisely the hypothesis of Rignano by means of which he proposes to explain memory, heredity, Lamarckian transmission, and all the teleological processes of organisms. I am deeply sensible of the pertinacity, the honesty, the thoroughness and the perspicacity with which Dr. Rignano has attacked these great problems, endeavouring to effect a truly synthetic solution of them in terms of one hypothesis. But I judge that he has not succeeded. I note first that in other expositions he finds himself driven to supplement this hypothesis with two others, namely, that of the universal tendency of organisms to maintain and restore 'physiological equilibrium' and that of a special form of energy that works teleologically. These seem to be confessions on his part of the inadequacy of the fundamental hypothesis set forth in the passage cited above. Secondly, I note that, like Semon's hypothesis of material memory-traces in the brain, the engrams of Rignano would, if we accepted his theory of their formation, account only for the reproduction of sense-impressions and complexes of sense-impressions, but

[1] *Upon the Inheritance of Acquired Characters* (Chicago, 1911).

not at all for the facts of true memory; for, as we have seen, the meaning of the sense-impressions is that which we chiefly remember and that which is of far more importance than the reproduction of sense-impressions. Further, true memory involves the placing of the remembered event in one's own past experience; and this also is over and above and altogether different from a mere reproduction, such as may be accounted for by the re-excitement of traces left in the brain by sense-impressions.

Thirdly, the hypothesis makes assumptions each of which seems highly questionable. It assumes that nervous processes consist in a flow of electric energy or of some form of energy closely analogous to electricity and that this energy may be of a very large number of specific varieties. Now physical science knows nothing of specific varieties of electric energy, beyond positive and negative charges. And the most delicate physiological research has failed to reveal any specific nervous energies. There is good evidence that some form of energy is liberated from the chemical latent state in the nerve-cells and transmitted from one to another; but there is no good evidence that this is of more than one variety. Further, if we admitted the possibility of a large number of specific nervous energies, it would remain very improbable that each of these could cause in the cells to which they were transmitted deposits of a substance (or as is probably meant a rearrangement of matter) such that it would upon occasion give out just the same specific variety of energy that had caused its formation. Rignano does not tell us whether he conceives the specificity of these assumed energies to consist in specificity of wave-length or frequency of a vibration of some sort. Yet this would seem to be the only possibility that is at all plausible; and no physical analogy can be suggested: red light does not specifically cause the deposit or formation of red substance, nor green of green substance, in any known instances.

As a final objection we may notice that the hypothesis is inconsistent with the other fundamental hypothesis by aid of which he and others seek to explain all the facts of seemingly teleological action, the hypothesis that all living matter has a tendency to maintain a constant condition of equilibrium and to restore that condition after every disturbance. For the facts of memory imply that every mental activity leaves a new trace of some kind; we cannot then attribute to the material structure of the brain two universal tendencies of *directly opposite mutually inconsistent kinds,* namely, the tendency to retain with the most marvellous faithfulness every change impressed upon it, and the tendency to restore itself immediately to its former condition after every disturbance.

Rignano's attempt to devise a physical basis for memory in the material structure of the brain fails utterly. Yet it is the work of one who is both physicist and biologist; and it has been made with greater thoroughness and knowledge and appreciation of the nature of the problem than any other.

I add here one other general consideration that adds to the improbability that memory can be founded in a purely material organization. It is a plausible view, a not improbable view, that in motor habits rendered perfectly automatic by much repetition we have the expression of a mechanism formed by the linking together of a number of nerve-cells by paths of low resistance. Those who pretend to identify habit and memory would regard memory as wholly conditioned by such linkages. But this, as Bergson has so clearly shown, is an impossible view. If, then, memory is based on material structure, it must be structural changes in the substance of the nerve-cells. Now, apart from the fact that the viscous matter of a nerve-cell seems remarkably ill-adapted to retain unchanged for periods of many years slight modifications of structure, we have no direct evidence of any such changes; and we have evidence that many nerve-cells subjected again and again to powerful excitations seem to undergo no change of structure, but rather to maintain their structure or to restore it rapidly after excitation to its prior condition. Thus the nerve-cells of the retina, which seem (so far as the most refined examination can show) not essentially different from other nerve-cells, continue to function in the same manner throughout life, although subject almost constantly during all waking hours to powerful stimulation. That is to say, in spite of constantly repeated and varied stimulation, these cells continue throughout the life of the organism to function in essentially the same way; they seem to retain no traces, their organization seems to undergo no modification through oft-repeated reception and transmission of impressions whether of similar or of widely varying kinds. And wherever we have direct evidence of the modes of functioning of nerve-cells we seem to find similar evidence of an unchanging monotony of function, a mere process of excitation and transmission of an unchanging kind.[1]

[1] I owe this argument to Mr. W. R. Bousfield's little book, *The Basis of Memory* (London, 1928).

NOTE 8. *The Legitimacy of Postulating Non-Spatial Factors in Vital Organization*

A dogma accepted by many philosophers asserts that we may not postulate any existents of a kind not experienced or given in experience. I find one of the clearest statements of this principle in the following passage: 'We think of mind as something over and above the continuance of enjoyments, and invent an entity superior both to things and to passing mental states. Such a mind is never experienced and does not enter, therefore, into the view of an empirical metaphysics. Nor is it of any avail to answer that, although not experienced, it must be postulated to account for certain experiences. The empirical method approves such postulation, which is habitual in science. But the unseen entities, atoms or ions which physics, for instance, postulates, or the molecules of the chemist, are all of them conceived on the analogy of something else which is known to experience. The mind, however, which is postulated in our case, is a mere name for something, we know not what, which claims all the advantages of the mind which we do experience, but accepts none of the restrictions of that mind, the most important of which is that it shall not go beyond what is found or suggested by experience.'[1] I do not find that any warrant is offered for this thesis. It seems to be a quite arbitrary dictum which is presented as an axiom. It has the advantage that it can be used to support almost any metaphysical system, the sceptical solipsism of Hume, the absolutism of Hegel or Bradley, or the neo-realism of Alexander. It permits of much latitude of interpretation. Much depends upon the slippery word 'experienced'. It is true that in forming our working hypotheses we should mould them after the analogy of others whose value has already been proved. But that seems to be as much as can safely be assumed.

In laying down this principle, Alexander would seem to mean, by 'experienced', perceived by aid of the senses. If this is the meaning of the dictum, it begs a very large question; and at the same time flies in the face of much successful scientific practice. As Alexander says: 'The

[1] *Space, Time and Deity*, by S. Alexander (London, 1920).

empirical method approves such postulation, which is habitual in science.' That is to say, science postulates whatever existents or modes of influence, whatever ultimates or indefinables, will best serve its purposes; and certainly does not confine itself to existents experienced in sense-perception. Whatever view we take of the ether, its postulation has undeniably proved useful; and the ether is not perceived or perceptible. A magnetic field is not perceptible. The postulation of potential energy has been fruitful; yet potential energy is imperceptible and it is not even modelled upon any perceptible existent.

If, on the other hand, 'experienced' does not mean 'experienced in sense-perception', the dictum loses all its restrictive power. If I hallucinate a ghost (or a centaur) or picture it in imagination, or merely talk about it, it is in some sense experienced. When a leading physicist tells us that perhaps matter pours into our universe from the fifth dimension through the great spiral nebulae, does he conform to this principle? Whence do philosophers acquire the authority to restrict his speculation? If certain physical phenomena cannot be explained without the postulation of an imperceptible fifth dimension, we must postulate it; and if certain organic phenomena cannot be explained without postulation of organization that is not spread out in three-dimensional space, we must postulate such organization.

There are indications that biologists are moving in the direction here indicated. A leading American biological chemist, Dr. R. S. Lillie, concludes a recent article on 'The Nature of the Vitalistic Dilemma' (*Journal of Philosophy*, 1926), with the following sentence: 'We may find it difficult to conceive of influences acting into the spatio-temporal continuum from without; but some such characterization of the conditions, seems required by many of the facts of vitality.'

Note 9. *On the Theory that all Matter is in Process of Dissipation*

In a recent essay on Cosmogony in the volume, *Evolution in the Light of Modern Knowledge* (London, 1925), a leading physicist, Sir J. J. Jeans, has presented the view of the history of the stellar universe that seems to him best supported by the evidence now available. It appears that it is impossible to make any plausible conjecture as to the origin of matter. 'The limit of our knowledge is that at present matter exists, or at least has enough of the attribute of existence to make us think it exists; as to what, if anything, it was before it was matter we know nothing.' In a later publication the same high authority suggests that, though we cannot conjecture the origin of matter, we may plausibly suppose it to have come into our physical universe from space of a higher dimension, each nebula being a sort of faucet or place of injection of matter into our universe from that other. He does not seem to take seriously Alexander's hypothesis that the matter of the nebulae emerged by condensation of that still more nebulous substance, Space-Time. But at least the evidence supports the view that the nebulae represent a relatively early stage of all material collocations. The matter within a nebula undergoes condensation by gravitation, giving rise to a sidereal system, such as our galactic system; within these systems planetary systems are formed by gravitational and tidal processes. But constantly accompanying all this condensation into smaller denser masses is a process of dissolution of matter into energy and its radiation into infinite space. Hence, unless there shall be continued formation of new matter or continued injection of matter into our physical universe, our material universe is destined to disappear by dissipation in the form of radiation. That at any rate seems to be the most probable fate of all existing matter.

Jeans sums up his account as follows: 'We have seen that all the stars fit on to a single evolutionary chain, and that the position of a star on this chain is determined by its mass. As a star gets older its mass decreases, the disappearing matter leaving the star in the form of radiation.

The end of a star, and indeed, so far as we can see, of the whole material universe, is simple—it is annihilation . . . in so far as anything is left behind, it will be intangible radiation travelling endlessly through space. The question inevitably suggests itself, whether this is perhaps only one side of the picture. We see radiation being generated out of matter every day of our lives, indeed every time we look towards the sun. Is there a converse process by which matter is generated out of radiation, so that the universe passes through endless cycles of births and deaths ? ' To this question, he says, it is impossible to give any answer or to discern any possibility of discovering an answer.

The best supported view is, then, that all matter is destined to annihilation by dissipation as radiation, and that all sidereal masses are constantly suffering such dissipation. Can we fairly call a process that consists in continual approximation to annihilation one of evolution ? Jeans does not scruple to do so ; his essay occupies the first place in a volume on Evolution. He speaks of the ' a single evolutionary chain ', on which all the stars have their positions. But this ' evolutionary chain ' leads only to annihilation. Can we properly call a continuing process of annihilation one of evolution ? I cannot see the propriety of such usage of the word. But the case is even worse : for, as Jeans tells us, it is possible that the converse process, the conversion of radiation into matter also occurs. It is true that he writes that, if this possibility is actual, ' the universe passes through endless cycles of births and deaths '. This suggests vaguely that, after each birth, the material universe undergoes for a period a process of evolution ending in death. But I judge that this passage about cycles of births and deaths is rhetorical only. We are offered no reason for supposing that, if radiation is converted into matter, this process sets in and goes on only when all matter has been annihilated. Rather it would seem that, if the conversion of matter into radiation is a reversible process, then changes in both directions are constantly going on in different parts of space ; and that all the matter of the universe is perpetually undergoing destruction and new matter being perpetually formed. This would be even less like a process of evolution than is a steady trend towards annihilation.

It seems clear that, when Jeans uses the word ' evolution ', he means, like so many men of science, nothing more than gradual change. In fact, in his second paragraph he writes of the ' conception of evolution or gradual change '.

It seems clear, then, that physical science can tell of no evolution of the stellar universe in any other sense than that it is the seat of gradual changes. And if we ask the same question of individual stars, we find only a similar answer. Jeans writes that the view he regards as best supported ' compels us to regard the evolution of a star as resulting primarily from a wastage of his mass '. But this wastage is merely the process of annihilation ; and the formation of planets and their satellites is merely incidental to this annihilation.

Can we then find any justification for speaking of the evolution of the parts of a sidereal system ? We are familiar with the expression, the Evolution of the Earth. It is often used, and it is the title of the second essay in the volume from which I have cited above.

When we turn to this essay we find merely the story of the condensation of gases, liquefaction and solidification, with the consequent formation of atmosphere, ocean and crust with its strata, its continents, mountains, etc. ; a phase of the annihilation process in which condensation is more prominent than dissipation. But that is all. And if we turn to particular masses of the earth's crust, say a mass of granite or a volcano, it is still the same story ; the mere rearrangement of varieties of matter, with some sorting out of the varieties by crystallization, gravity, tidal and river action and so forth. There is nothing of the nature of evolution in all this. We arbitrarily single out some chunk of matter, a volcanic cone, a piece of granite or basalt, a bed of sand or mud ; we describe the changes by which it has been brought together in its present form ; and we call it an evolution. But if ' evolution ' means nothing more than gradual change, change that is not instantaneous, we shall do better to resign the use of the word and stick to the word ' change '.

NOTE 10. *Creation by Evolution*

It is perhaps significant that in the recently published volume *Creation by Evolution*,[1] to which twenty-six high authorities contribute an essay apiece (each not at all concerned with the how of evolution but wholly devoted to proving that evolution has occurred) only a few pages are concerned with evolution in the physical realm. These passages occur in the last of the essays (written by Dr. H. H. Newman). The author begins by asserting : 'No greater mistake about evolution could well be made than to limit its application to living organisms. There has undoubtedly been as real an evolution of the cosmos, of the solar systems, of the earth and other planets, of the molecules, and of the atoms as there has been of organisms.' This is the proposition which I here traverse and would utterly deny. I submit that the enunciation of it implies a total lack of appreciation of the essential nature of evolution and of the profound difference between the evolutionary changes of the organic realm and the mere changes of the inorganic.

When we turn to Dr. Newman's account of evolution in the physical realm, what do we find ? A statement of the alleged evolution of stars, of planets and of atoms. We are told that stars in general run a course of increasing condensation, ' giving off enormous amounts of energy '. Occasionally a star comes near another sun, and then any family of planets it may have ' will be broken up and another one born. This rhythm may go on for ever, so far as we can tell, for there appear to be no agencies tending to put an end to it.' He should perhaps have added that two stars may collide with consequent renewal of what, in virtue of a false analogy, is called the youth of the system and a repetition of the process of condensation. Is such a ' rhythm ', such alternation of condensation and rarefaction an evolution ? Surely not!

As regards individual planets the evidence of evolution adduced is that ' since the earth reached its present diameter . . . there have been no less than half a dozen major pulse beats of the earth ; and numerous minor rhythmic changes have been superimposed upon the major rhythm '. Here

[1] London, 1927.

again an alternation of changes of opposite kinds and nothing more; nothing that can in any proper sense be called evolution. The illegitimate use of the word ' rhythm' merely serves to suggest vaguely some analogy between the planet and organisms; for, in the latter, rhythmic pulsation is a common phenomenon.

As regards the atom, the planetary theory of it is accepted (which we are now told in other quarters is quite out of date). We are then told of the hydrogen atom that it is the simplest of all. ' Other atoms are far more complex than the hydrogen atom, some of them containing over two hundred times as many protons and electrons. No matter how large and complex an atom *becomes* [italics mine] it includes no other kinds of particles than those contained in the simplest atom. All differences in the properties of the elements are due to the number of and the variations in the arrangement and configuration of these ultimate particles.' Here it is implied that the hydrogen atoms have evolved into more complex atoms. But is there any positive evidence of such a process? None is mentioned; but in the following paragraph we are told how the more complex atoms become less complex. ' There is a sort of unrest in the nucleus which operates to break down the equilibrium existing among the protons and electrons and results in the shooting off . . . of electrons and of groups of protons and electrons from the nucleus. . . . The emission of these rays is nothing more or less than a process of evolution of elements, one element becoming transformed into another. By radio-active disintegration the most complex elements, such as uranium, thorium, and radium, are reduced slowly and by distinct steps to elements of less complexity and greater stability.' The only transformation described as actually occurring is a disintegration of complex atoms; and this our author calmly calls ' a process of evolution '.

But suppose that the reverse process, the change from less to greater complexity of atoms, also does occur; then an atom under certain conditions may become more complex, and again under other conditions less complex. Does such a ' rhythm ' constitute a process of evolution? To give that name to such an alternation is merely an abuse of language.

It is of interest to note in the same volume that not all emergent evolutionists assert that emergent evolution has occurred in the physical realm. Thus Dr. H. F. Osborn, who contributes the Foreword to the volume, writes: ' Evolution is the incessant appearance of new qualities, new characters, new powers, new beauties, for which there is no antecedent in experience or no evident promise in the germ itself.' Further: 'This originative and creative principle of emergence, of creative evolution, appears to be lacking in the lifeless universe, even as revealed by the recent and most marvellous discoveries in physics and chemistry, and in astronomy are not new physical elements compounded by the simplification or complication of older physical elements, to give rise to new forms, but without the creation of new forces ? Is there not invariably in the physical and material world antecedence and consequence, cause and effect ? Are we not, therefore, facing in the biological world a new recognition of the order of Nature in the incessant creative, emergent evolution of new forms, of new characteristics, of new powers ? Consequently the addition of new powers and new properties seems peculiarly distinctive of life.'

It is not clear whether Dr. Osborn would agree with Dr. Newman and Herbert Spencer in asserting evolution of some kind in the physical realm ; it is clear only that he denies it emergent evolution. This, of course, is to spoil the emergent scheme from the point of view of the thorough-going emergentist, in whose eyes its chief value is that it seems to bridge the gap between the physical and the vital realms.

Again, in the same volume the first essay by Dr. David Starr Jordan is entitled, ' Evolution—Its Meaning '. He begins by asserting: ' By evolution, as the word is now used, we mean the universal process of orderly change. It includes cosmic changes in suns and planets and organic changes in living creatures. . . . And from the fact that all these changes . . . are orderly, never random nor accidental, we derive our definition of evolution. Moreover, as this process occurs throughout all that we know, evolution becomes another name for Nature.' Every natural process or event is thus brought under the term ' evolution ', and the term loses all special meaning and value ; it comes to mean any and every process in nature—the fall of a stone or of a raindrop, the dance of dust particles in the breeze, the explosion of gunpowder, or the destruction of a tree by a stroke of lightning, all alike are processes of evolution. Having thus by definition at once extended the meaning of the word evolution to cover all natural events without exception, having made the word applicable to the physical realm at the cost of rendering it quite useless, Dr. Jordan hastily leaves the topic of evolution in general (after devoting half a page to it) and devotes the rest of his essay to organic evolution. He proceeds : ' Evolution, indeed, is Nature's way ; . . . That Nature has her ways is the most visibly evident fact in all our experience, and such phrases as ' blind force ' have no real meaning. Nevertheless, the forces and conditions which surround suns and planets, or which mould mountains and seas, or which determine the formation of crystals or the accumulation of rocks, differ in certain ways from those which modify generations of life. We therefore usually treat orderly change in organized beings under a special head, that of organic evolution.'

NOTE II. *The Causal Efficacy of Emergent Mind*

In view of the fact that Lloyd Morgan is one of the most influential exponents of Emergent Evolution and that his statements concerning the emergence of Mind are ambiguous or positively contradictory, it is of some importance to show that, as widely understood and accepted, Emergent Evolution, which professes to describe the emergence of Mind from a purely physical world (that is to say, from matter or some forerunner of matter, from physical reality as mindless to all appearance, even to the most minute investigation, as the inorganic matter about us, from events as purely mechanistic as the physical events dealt with by astronomical or optical science), regards mental events as causally efficacious, as affecting the course of the brain-processes, from which they emerge. This is clearly asserted by Lloyd Morgan in the general principle that all emergents are effectively related to the events from which they emerge, make a difference to the ' go ' of those events, as also in his many assertions of effective conscious guidance of behaviour.

One of the first expositions of Emergent Evolution is Dr. L. T. Hobhouse's *Development and Purpose*,[1] a masterly discussion of evolution, prepared for by his earlier *Mind in Evolution*. The principles of configuration and of emergence are clearly set forth, and Life and Mind are conceived as having emerged from the inorganic or physical realm; yet, having emerged, the psychical is not regarded as ineffective in the life of the organism. Conscious activity is regarded as having causal efficacy in the organic whole, and as working purposively or teleologically. ' An organic unity,' he writes, ' is harmonic and teleological. If that is so, the organic process proper rests on a causation that differs in kind from the mechanical. By perfection of predetermined arrangement the mechanical may acquire more and more of organic and purposive character, but in the limit, where the correlation is complete, it passes over into the region at once of organicity and of purpose.' That seems to mean, not a sudden emergence of teleological causation, but a step-by-step emergence by which mechanistic process is more and more modified until the

[1] London, 1913.

events in the organism become distinctively teleological. Again: 'Purposive activity, i.e. the conditioning of the action of each part of a system by the causal tendency of the configuration as a whole, is the characteristic mode of reaction of certain structures—those which we call psycho-physical. *Qua* physical this structure tends to act in accordance with mechanical laws, but this action is modified by the condition mentioned, which is the psychical element of the whole in operation. . . . Every element of living tissue capable of genuine organic self-adaptation is affected, along with its mechanical interactions, by a psychic factor operating, whether with the rudimentary or fully developed character of a purpose.' Mind, then, having emerged has effective teleological causal efficacy. And here is Hobhouse's explicit formulation of the emergence of Mind and consciousness: 'Applying to Mind the general considerations as to development, which have been explained, we regard it essentially as a mode of activity dependent for its specific character on the co-operation of elements. These elements, as long as they exist apart, would not constitute the peculiar form of unity which is Mind, but would be related to it as the chemical molecules which constitute a cell are related to the living cell. If these molecules come together to form a cell, they undergo some development either by the unlocking, or by a mutual modification in the action, of pre-existing energies, and if that cell is conscious, the pre-existing energies must be conceived as containing or exerting activities which unite to form the activities of consciousness, just as they exert pressures and tensions which in combination yield the phenomena of contractility."

This is, I think, the most thorough-going of all statements of the emergence of Mind, of causally efficacious purposive intelligent activity.

Dr. W. M. Wheeler, in his brief exposition of *Emergent Evolution*,[1] defines 'emergence' in words that leave no doubt that he also means to accept the emergence of causally efficacious Mind. '" Emergence ", in the following pages, signifies neither the manifestation or unveiling of something hidden and already existing, as in the common and the entomological denotations of the word, nor some miraculous change . . ., but a *novelty of behaviour* arising from the specific interaction or organization of a number of elements, whether inorganic, organic or mental, which

[1] New York, 1928.

thereby constitute a whole, as distinguished from their mere sum, or resultant.'

Mr. A. D. Broad [1] writes benevolently of Emergent Evolution, and might almost be regarded as one of its exponents. But it is impossible to pin him down as accepting any view or theory. The most that can be said is that his statements of the principle of emergence would, if he accepted Emergent Evolution and the emergence of Mind, seem to imply that he would accept it in the thorough-going sense of Hobhouse and Wheeler. However, he never gets close to the problem of teleological causation; he never even formulates it as a possibility. In his discussions of teleology he is concerned only with teleological systems in the sense of mechanical structures that seem to have been designed, i.e. he is concerned only with what was defined in the text as pseudo-teleology or derived teleology. On the question of the relation of Mind to matter he reaches the conclusion: 'I judge the most likely to be some form of the Compound Theory which is compatible with Emergent Materialism.' And by this he seems to mean that Life and Mind have emerged from a physical world that had nothing of the nature of Mind; but that Mind having emerged involves a 'Psychic Factor'. This Psychic Factor may be material; but, if so, it is very different from ordinary matter. 'It is not destroyed by the breaking up of the body with which it is connected; it does not manifest itself to sense-perception; and it does not produce ordinary physical and chemical effects.' Yet, by implication, the Psychic Factor, having emerged, is capable of making a difference to the 'go' of organic events.

Professor S. Alexander also (like Mr. Broad) is more concerned to state what things are or are made of than with the question of how events take place, and it is difficult to find any passage in which the causal efficacy of mental events is explicitly asserted; but it is implied throughout his exposition of Emergent Evolution (in *Space, Time and Deity*). We are told that mind is made up of conations and that 'the mental act is a conation, which is something mental, and not merely physiological'. Also, 'Primarily conation is practical, and it issues in movements which tend to alter or destroy the object or at least to affect our relation to the object.' And 'In mind mental acts are also connected causally with one another, and the mind is subject

[1] *Mind and Its Place in Nature* (London, 1925).

to determination like all other things.' And 'although mind cannot be and act without things from which to select its objects, neither the things nor the objects are affected in themselves by the presnce of mind except so far as the mental conation alters them.'

Professor H. S. Jennings,[1] jubilantly welcoming Emergent Evolution, writes: 'If a student of humanity asserts that man shows certain characteristics, his assertion is not negatived by the fact that no such characteristics are to be found in other organisms. There is no *a priori* ground for sneering at the notion that man in some respects acts on principles diverse from other animals. *A priori* principles of this sort don't go, if Emergent Evolution is a correct doctrine; such questions are purely matters for investigation.' Again: 'Mechanical science asserts, not only that all action is determined before it occurs, but that it is determined by the physical conditions, by the material situation; that is, it teaches materialism, with all its gross consequences. . . . It cannot be denied that men speaking in the name of biological science have proclaimed the basic doctrines from which those conclusions have been drawn; from which they are perhaps justly drawn. The immutability of the laws of nature, the theoretical predictability of the future from the past, the denial that anything essentially new can occur—these are almost commonplaces of the schools. The explicability of all that occurs in the living through a knowledge of the laws of the non-living is a dogma in wide circles of biology. The incompetence of the mental to affect physical happenings has become a widely held doctrine, urged by biologists, philosophers and psychologists. The objective examination of behaviour, we are told, leaves no role for the psychic; as one physiologist expressed it "the sensations, memory, thoughts disappeared like fluttering forms of vapour,nowhere remained the smallest spot for the psyche". Neal felt that he was opposing a generally accepted scientific doctrine when he had the hardihood to maintain that consciousness makes a difference to what happens. So nearly a commonplace has become the doctrine of the inefficacy of the mental that one finds a writer in a philosophical magazine raising and apologetically, with the fear of the biologist in his soul, the questions whether it *may not* after all be true that "purpose makes a difference" and that "intelligence is practical and a

[1] *Science*, January 14, 1927.

source of power". So low has the perfect doctrine of mechanism brought us!' He goes on to say that Emergent Evolution does away with all these dogmatic denials. 'According therefore to radical experimentalism, consciousness does make a difference to what happens; particular types of consciousness make a difference. Emergent Evolution asserts this from another point of view; the conscious emergent is one that acts on different principles from the unconscious one. Emergent Evolution so does away with that monstrous absurdity that has so long been a reproach to biological science; the doctrine that ideas, ideals, purposes have no effect on behaviour. The mental determines what happens as does any other determiner. . . . The desires and aspirations of humanity are determiners in the operation of the universe on the same footing with physical determiners.'

Professor C. J. Herrick is no less explicit in asserting the causal efficacy of emergent mental events. In two recent articles he has made the assertion in no uncertain terms, e.g.: 'The prevision of possible future consequences of action is a real causative factor in determining which course of action will actually be chosen.'[1] And 'We regard conscious experience as a part of the vital process . . . as a causative factor in human behaviour . . . the fact that in the case of the man there is some awareness of what is going on is by no means a negligible phenomenon; for, as we have seen, this awareness is itself a causative factor in the process.'[2] These and many other passages leave no room to doubt that, in the view of this distinguished neurologist, conscious processes are causally efficacious and affect the course of the brain-process; though it is equally made clear that in the same view their causal efficacy is of the mechanistic type.

It would be possible to add to the foregoing list. But it will suffice to show that a change of opinion on this matter is in progress. During thirty years my advocacy of this view now coming into favour has made me a pariah among men of science; and I naturally note the change with pleasure.

[1] 'The Natural History of Purpose', *Psychological Review*, 1925.
[2] 'Biological Determinism and Human Freedom', *International Journal of Ethics*, 1926.

NOTE 12. *Various Versions of Emergent Evolution*

It has been widely felt that one of the most unsatisfactory features of Herbert Spencer's scheme of universal evolution was his account of the appearance of Mind on the cosmic stage. Animals were evolved, nervous systems and their mechanical reflex processes increased in complexity; and, when some unspecified degree of complexity was reached, consciousness appeared and Mind began its futile course as a helpless spectator of cosmic events.

The chief purpose of Emergent Evolution is to amend this defect in the evolutionary scheme; its chief claim to superiority over the older doctrine is its claim to render intelligible the evolution of Life and Mind from the inorganic and to rescue Mind from its position as a mere spectator, a position in which it enjoyed *otium sine dignitate*. But no one of the several schemes of Emergent Evolution hitherto elaborated carries through the programme in uncompromising fashion. A thorough-going uncompromising scheme of Emergent Evolution would describe evolution as beginning from some physical universe in which there was no trace or shadow or gleam of any of the qualities or modes of happening that characterize living organisms and mental events, a realm of purely mechanistic events. And it would describe, and seek in some sense to explain, the evolution by emergence of increasingly complex molecules and material structures and mechanistic events, followed by emergence of the characteristic functions of living things and of Mind; each later and higher emergent being effectively related to those from which it emerged, that is to say, exercising causal efficacy, making a difference to the general course of the events or systems of events within which it was emergent.

Although many of those who accept Emergent Evolution seem to imply that such a scheme is what they accept, it would seem that no such scheme has been elaborated. In each scheme offered us there is compromise or vacillation. Something of Life or of Mind is imported into the physical universe from which Mind is said to have emerged.

From the earliest times cosmological speculation has generally regarded the physical realm as in some sense and degree controlled by Mind; either by Mind extrinsic to itself, the mind of superhuman anthropomorphic beings, or by Mind or mind-like powers intrinsic to matter. These two views have usually been alternatives; either one renders the other unnecessary. The former has usually been the choice of the religiously minded; the latter of those who aimed at a strictly scientific cosmology. Even the champions of strict Materialism have for the most part attributed something of Mind to matter. Lucretius wrote of the declination of the atoms; La Mettrie, Büchner and Ernst Haeckel, and many other such loose-thinking exponents of Materialism, have adopted what we may call 'intrinsic anthropomorphism' in their endeavours to make plausible the view that man is *nothing but* matter. Haeckel's statements are typical, as for example: 'All natural bodies known to us are equally animated . . . the distinction which has been made between animate and inanimate does not exist.' And again: 'Liking and disliking, desire and aversion, must be common to all atoms, which can only be explained by ascribing to them perception and volition.' But modern science has prided itself on getting rid of anthropomorphism, both extrinsic and intrinsic. It is, then, surely not without significance that all those contemporary thinkers who have elaborated schemes of Emergent Evolution, although their schemes are put forward, not as loose popular expositions, but as made with strict regard to the principles and teachings of modern science, have not avoided anthropomorphism either extrinsic or intrinsic, or both together; that is to say, they have imported something of the nature of Mind as we know it in ourselves into their picture of the physical universe.

At first sight, Dr. L. T. Hobhouse's scheme, which is the most thorough-going in that it fully and explicitly recognizes that mental activity is both teleological and causally efficacious in the guidance of behaviour, seems to be an exception. From the first edition [1] I get the impression that he is describing the emergence of teleological mental activity out of a purely mechanistic physical world. The principle of configuration is clearly stated. Teleological causation is accepted as characteristic of organisms. And we are told that 'In proportion as mechanical adjustment becomes comprehensive and immediate it approaches the organic character. It could reach it only at the point where

[1] *Development and Purpose* (London, 1913).

the succession of cause and effect, as between part and part, merges into the simultaneity of a consentaneous whole. . . . We conceived two separately centred but rapidly interweaving processes, and we could increase the rapidity and closeness of interaction at pleasure. As we do so the action of each part is more and more closely determined by the consequent action of the residue. At the limit it is so determined from the first, i.e. the action of the part is as such conditioned by actions of the residue which it itself sets up. The limiting, correcting activity of this residue is now operative in the actions of the part itself from the first. But this is as much as to say that the action of any part of an organic whole is action conditioned by its result, i.e. is teleological. . . . A system of parts so related that each is thus conditioned in its action by its effect in maintaining a residue is a harmonious system. An organic unity then is harmonic and teleological. If that is so, the organic process proper rests on a causation that differs in kind from the mechanical. By perfection of predetermined arrangement the mechanical may acquire more and more of organic and purposive character, but in the limit where the correlation is complete, it passes over into the region at once of organicity and of purpose.' This seems to be unmistakably an attempt to exhibit teleological causation as emergent within mechanistic systems. Further: 'Applying to Mind the general considerations as to Development . . . we regard it essentially as a mode of activity dependent for its specific character on the co-operation of elements. These elements, as long as they exist apart, would not constitute the peculiar form of unity which is Mind, but would be related to it as the chemical molecules which constitute a cell are related to the living cell. If these molecules come together to form a cell, they undergo some development either by the unlocking, or by a mutual modification in the action, of pre-existing energies, and if that cell is conscious, the pre-existing energies must be conceived as containing or exerting activities which unite to form the activities of consciousness, just as they exert pressures and tensions which in combination yield the phenomena of contractility—that is to say, Mind in the organism is not to be conceived as either external or as growing out of something like matter, taken as wholly discrepant from it. It is to be conceived as a synthesis of elements, which do not function except in combination.'

And we are told that if we compare the successive stages of evolution 'we find evidence of Mind emerging under conditions of which by degrees it becomes master'.

These and many other passages seem to justify the statement that Hobhouse is attempting to describe the emergence of Mind from the physical realm. But in the new edition we learn that 'Mind is not all reality nor all that moves or exercises control. Fundamentally it is an element in reality', and 'an element coeval with the rest and an essential condition of their existence. A world without Mind is therefore an abstraction, not a real condition of things. . . . The structure and process of Reality as we find it is determined by the effort of Mind towards harmony among elements which but for it are mutually indifferent.' Hence, 'On our theory we should not expect to find life originating from inanimate matter. . . . We infer a Mind at the root of the structure and more especially of the development of Reality, functioning, that is, in Reality at large and functioning as a unity.' These and similar passages seem to teach that Mind has not emerged from the physical realm; they seem clearly to imply a dualistic view similar to that advocated in these pages, under which teleologically working Mind is not in any sense derivative from the mechanistic physical realm. They seem to acknowledge that the attempt made in the first edition was not successful. Hobhouse abandons the attempt after making a resolute but unsuccessful effort.

Four other schemes of Emergent Evolution claim our attention, those of Lloyd Morgan, of Strong, of Noble, and of Alexander. The argument of this volume requires a brief examination of each of these.

Mr. Lloyd Morgan's Version

Lloyd Morgan's effort is notable as giving us what seems to be a thoroughgoing scheme of Emergent Evolution, and yet as complicating it with a form of intrinsic anthropomorphism quite inconsistent with it, and an extrinsic anthropomorphism that seems otiose.

We are told that there has been evolution by emergence in the physical realm; the properties of matter have emerged from some unspecified forerunner; those of the complex chemical compounds have emerged from the conjunction of simpler chemical compounds; properties [1] distinctive of living things have emerged from these conjunctions of these complex molecules; and Mind has emerged from living things. In all cases emergence has occurred when new conjunctions occurred, conjunctions bringing about new kinds of relatedness; and the emergent properties are in each case themselves effectively related to the systems from which they emerge, that is to say, they have causal efficacy, make a difference to the go of the system. 'In order that there should be a difference in the course of events the relatedness in question must be what I call effective. By this I mean that when it is present some change in the existing go of events occurs, which would not occur if it were absent.' Again: 'When some new kind of relatedness is supervenient (say, at the level of life), the way in which the physical events which are involved run their course is different in virtue of their presence—different from what it would have been if life had been absent.' Again: 'The new relations emergent at each higher level guide and sustain the course of events distinctive of that level.'

From our point of view it is especially importaut to note that this is clearly maintained also of the mental emergents, as in the following passages and many others: 'I shall have occasion hereafter to urge, as against radical behaviourists, that mental guidance of events counts for progress and betokens a kind of relatedness that is effective. When it is present changes occur which do not occur in its absence. The manner of go in the enriched system is different. That is what I mean by speaking of guidance as dependent on the supervenient kind of relatedness at the level of Mind.' 'In a physical system wherein life has emerged, the way things happen is raised to a higher plane. In an organism within which consciousness is emergent a new course of events depends on its presence. In a person in whom reflective thought is emergent behaviour is sustained at a higher level. . . . At the level of contemplative thought, how perception runs its course depends on the guidance of reflective consciousness, so far as co-existent; and how what is given in sensory presentation takes form depends on the guidance of perception, if that level have been reached.'

[1] Lloyd Morgan distinguishes qualities intrinsic to things from the properties they manifest in relation with other things. I use the one term 'property' to cover both.

Consciousness, then (as also 'cognitive relatedness') emerges 'as something genuinely new, at a critical stage of evolutionary advance'; and, when it has emerged, it guides the course of events in the brain, it has causal efficacy in the complex psycho-physical events that issue in bodily behaviour. And unreflective consciousness also ('that on which the guidance of animal behaviour in large measure depends') is similarly effective, has causal efficacy of some lowly degree.

This recognition of the causal efficacy of conscious cognition in the guidance of events within the organism that issue in behaviour is the most important and distinctive feature of 'Emergent Evolution'. It involves the definite rejection of Epiphenomenalism and of Psychol-physical Parallelism, and ascribes to conscious thinking, or experience as such, a role, a function, an effective part, in short, causal efficacy, in human action and in the drama of evolution. It involves recognition of interaction between the psychical and the physical. For consciousness guides the course of brain events ; and reflective consciousness is surely psychical; while the brain events are physical processes raised to the vital plane by the presence and co-operation of vital emergents.

Lloyd Morgan's oft-repeated assertions that Life and Mind are emergents and that emergents are effectively related to the system from which they emerge ; his frequent references to conscious guidance, his distinction, often repeated, between, on the one hand, bare sentience and enjoyment which do not guide behaviour, and, on the other, reference, prospective reference, and foretaste in enjoyment, which do bring 'conscious guidance with prevision'; his insistence on 'cognitive reference, which, as prospective, subserves some small measure of conscious guidance of the course of events which we observe', i.e. the events of animal behaviour ; all these seem to be assertions, as explicit as possible, that mental or conscious events emerge in the same sense as the properties that characterize Life emerge, and as those that characterize chemical compounds emerge ; and that these conscious events, having emerged, exert influence upon the course of events in the systems from which they are emergent. In a score of passages his language will bear no other interpretation.[1]

[1] The passage most explicit in this respect is perhaps the following. Discussing the status in the evolutionary series of the behaviour of the

Yet running all through this scheme of thoroughgoing Emergent Evolution, or, rather, interleaved with it, is a doctrine strictly inconsistent with it, namely, the doctrine of Spinoza that the physical and the psychical are two attributes of all reality, two attributes or aspects that cannot be regarded as in any sense influencing one another.

This doctrine Lloyd Morgan speaks of as one of universal correlation or concomitance of the physical and the psychical. 'We also acknowledge unrestricted correlation of the kind Spinoza postulated under his doctrine of attributes. Within the domain of both attributes there is continuous development under progressive emergence. Each ascending stage in the one attribute is evolved with that of the other. Neither is evolved from the other.' And: 'Mind is within one of the two attributes of nature. It is the natural correlate of certain physical events which belong to the other attribute. There is, for us, no effluence from either attribute to the other; nor is there any causal influence of the one on the other.' Again: 'There are no physical systems of integral status that are not also psychical systems; and no psychical systems that are not also physical systems. All systems of events are in their degree, psycho-physical. Both attributes, inseparable in essence, are pervasive throughout the universe of natural entities.'

These two inconsistent doctrines are, I say, interleaved rather than combined; for they cannot be logically combined.[1] Yet in some passages the attempt is made. Thus

just-hatched chick, and using the word 'behaviourist' to characterize that type of interpretation of behaviour which denies all influence of mental factors and explains all behaviour as mechanical reflexes or physically caused responses to stimuli, he writes: 'On a basis of correlation one has to distinguish a *primitive psychical system* before the quality of consciousness . . . emerges, from a *primitive mind* in which it *is* emergent. . . . My interpretation of the chick's status is frankly behaviouristic, if a correlated psychical system not yet effective in guidance be acknowledged. But *pari passu* with the evolution of its behaviour there is developed proficient reference to that towards which it behaves. And with this comes conscious guidance, which the behaviourist on his part will not allow.'

[1] The most explicit instance of interleaving the two inconsistent doctrines is perhaps to be found on page 37 of *Life, Mind and Spirit*. Here the author insists on the separateness of the stories of the two aspects: '*as stories* neither can interact on the other; "this" aspect of the natural events themselves can in no wise interact with "that" aspect'. In the next paragraph he writes: 'It may then be asked: How, on this hypothesis, can mental reference count in the guidance of events, as it assuredly does in many forms of animal behaviour and in all forms of human conduct? The answer we shall give turns on

he writes of the pyramid of Emergent Evolution, material in its lower parts, vital and mental in the upper levels. ' In our pyramid of Emergent Evolution, the ultimate basis under such acknowledgment is a world of purely physical events (and their correlates) in changing spatial and temporal relatedness.' Here in one sentence is the clearest expression of Lloyd Morgan's desire to eat his cake and have it. The physical events from which Mind is said to emerge are said to be *purely* physical and yet they are said to have (in brackets) their psychical correlates. And when, at an advanced stage of evolution, the psychical correlates emerge from their brackets and enter into effective relatedness with their physical correlates, they are said to have emerged from the *purely* physical events and to exert conscious guidance on them and yet to be without causal influence upon them.

This major inconsistency within Lloyd Morgan's scheme overshadows all others. But it is worth while to insist on a minor one.

In the earlier of the two volumes of Gifford Lectures, the evolution of ' reference ' or cognition by way of the compounding of sense-qualities, immediately given and revived, is painstakingly elaborated. In the second volume, this elaborately supported claim to have evolved Mind from that which is not Mind is allowed to lapse. In place of it we are given an equally unsuccessful attempt to evolve foresight or prospective reference from bare objective reference, which in this volume takes its place among the unevolved or primordial psychical correlates of the primordial physical events. And conscious guidance, which in the earlier volume begins with the emergence of reference, in the second volume becomes the peculiar privilege of ' prospective reference with foretaste in enjoyment'.

Less serious perhaps as a defect of Lloyd Morgan's scheme of Emergent Evolution is the fact that he has a third string to his bow. Not content with the causal efficacy of his emergents, Life and Mind, for the explanation of evolution,

the evolutionary story of reference when some *prevision* of that which is coming is concomitant with some *present* mode of action in the life-story, implying influence received under external stimulation and given back in behaviour.' In this way he lays down for himself the programme of showing how what he has declared to be impossible (according to his fundamental assumption of universal correlation), namely, psycho-physical interaction, is rendered possible by the principle of emergence.

he asks 'What makes emergents emerge? His answer is Causality. And by Causality, as something distinct from the universal causal efficacy of physical events and their emergents, he means the directive Activity of Divine Power. He tells us that, in spite of the effective relatedness of emergents expressed in 'Causation throughout the effective field of the universe' (which he says 'is present at all times and in all places') his scheme of Emergent Evolution remains merely descriptive until this Activity is acknowledged. And this acknowledged Activity, which is omnipresent throughout Emergent Evolution from top to bottom, is not, then, like the Activity of the Deity of the Deists, the creating of a mechanism which, once set going, runs its predestined course without further guidance. But, like that Activity, it is admittedly teleological. It is a planful Activity, the nature of which we conceive only in the light of our own experience of activity; and it is the expression of Divine Purpose.

Now, I am not quarrelling with Lloyd Morgan's postulation or acknowledgment of Divine Purpose and its expression in teleological direction of the course of evolution. I merely point out that, while he nowhere explicitly recognizes the essentially teleological nature of human action and leaves us uncertain whether in his view our teleological mental guidance of action does or does not emerge from the realm of mechanistic causation, he nevertheless does not hesitate to postulate teleological Activity as real, fundamental and unevolved. He strains at the gnat and swallows the camel.

I reserve another objection to Lloyd Morgan's scheme for statement after brief consideration of the closely allied scheme of Professors C. A. Strong and Durant Drake and R. W. Sellars.

Prof. C. A. Strong's Version.

This scheme makes, as its fundamental assumption, one similar to Spinoza's and Lloyd Morgan's assumption of universal concomitance or correlation.[1] But though similar it is different, and differs in a way that enables this

[1] This may fairly be called an improved form of Spinoza's doctrine. It has been propounded more or less independently by the late Professor Paulsen and by Dr. Morton-Prince, and most thoroughly by C. A. Strong in his *Why the Mind Has a Body* and his *Origin of Consciousness* (New York, 1918).

school of thinkers to avoid the principal inconsistency of Lloyd Morgan's scheme. While the latter is a generalized form of Psycho-physical Parallelism, the former assumes, not that the physical and the psychical events run parallel as two attributes of some inconceivable reality, but that all reality is psychical and that all events we call physical are the appearance to us of events which in reality are psychical. If Emergent Evolution makes this assumption as regards the intimate nature of all reality, it can legitimately assume the successive emergence of more highly organized forms of Mind, with, at each level, emergents effectively related to the systems from which they emerge. It can without gross inconsistency assert: 'Certainly all animals, including ourselves, are made of the same stuff that makes up the rest of the world'; [1] and that 'animal behaviour, including our own, is merely a complex resultant [or emergent] of the behaviour of these ordinary atoms in their new and peculiar situations'; and yet can say: 'Certainly purpose, desire, will are efficacious and important'; and can speak of 'a theory like ours, which makes not only will-events but every mental event causally efficacious'.

The exponents of this view have not yet fully assimilated the emergent principle; but there is no reason why they should not do so, and it is difficult to see why Lloyd Morgan has chosen the Spinozistic assumption, which is not compatible with his Emergent Evolution, rather than this other assumption, no more difficult in itself and infinitely more capable of being combined with a thorough-going Emergent Evolution.

Dr. Strong has, so far as I can find, nowhere grappled with the problem of the relation between mechanistic and teleological events. He has, like so many others, concentrated his attention on the question: What stuff are things made of? to the neglect of the more important question: How do events run their course? We are left with the impression that, for all his insistence on the psychic nature of the stuff of all things, all events are for him mechanistic.

Dr. Drake, the colleague of Dr. Strong in the elaboration of this scheme, asks: 'Is this view mechanistic?' He replies: 'Not *necessarily*, in the narrow sense.' He proceeds to argue that cerebral processes seem on the whole

[1] This and following citations are from *Mind and Its Place in Nature*, by Durant Drake, a clear and concise presentation of this variety of Emergent Evolution.

to be regular and *therefore* mechanistic. He shows clearly that for him the meaning of 'mechanistic' and 'non-mechanistic' is a matter of regularity and irregularity. He writes: 'In any case it is the cerebral processes whose regularity or irregularity is in question; and if they do seem, on careful study, to be hopelessly irregular in spots, we can use the term "vital principle", or anything else you please, to point to the irregularities.' And, although he admits the causal efficacy of desire and will and of all mental events, he conceives them to work mechanistically, simply because he has not grasped the essential nature of teleological or purposive action, but regards it as merely some utterly irregular non-natural or miraculous interference with mechanistic events. Purpose, he says, is merely the working of a self-righting mechanism. Having, then, failed to conceive of teleological causation as anything but a miraculous interference with the course of nature, this school naturally does not attempt to expound the emergence of teleological mental events from the mechanistic psychical realm which appears to us as physical events. It makes easy the task of Emergent Evolution by asserting that all matter is psychical; but it neglects to take account of the most distinctive feature of the psychical, namely, its teleological functioning; and only by such neglect does it make plausible its description of matter as essentially psychical. It remains then a self-consistent scheme, but at the cost of ignoring the main problem.

Objections applicable to both Lloyd Morgan's and Strong's versions of Emergent Evolution

The versions of Emergent Evolution propounded by Lloyd Morgan and by Strong are alike in one respect which is, I argue, a very serious defect. Both postulate that the realm of things commonly called physical is real and objective and that in its real nature it is psychical (Strong) or has a psychical attribute (Morgan). Now this assumption of the psychical nature of the physical realm is made purely and solely for the purpose of avoiding the difficult assumption that psychical reality has emerged from the purely physical or material. There is no other ground for it; we have no spark of evidence that supports it, no faintest indication of any type of physical event that requires any such hypothesis for its explanation; the hypothesis does not, as any legitimate hypothesis must (and as the hypothesis of purposive activity in the animals does) serve in any way to guide us in the making of observations, observations that would serve to prove the hypothesis.

Again, it may be asked: Is the process by which these authors are led to conceive the physical as psychical a legitimate one? Consider it. We conceive the psychical and distinguish it from the physical only through inspection of our conscious activities. We then strip away from the psychical its every distinguishing mark; and we have left a bare word empty of all meaning except the purely negative meaning of not-physical; the psychical thus defined is neither feeling, nor cognition, nor conation; it is not even bare sentience; it is not teleological events or activity; it has no qualities, properties, relations, substance or shadow; it has no assignable nature; it has no causal efficacy; it is nothing. We then go on to say that this vague empty not-physical is identical with the physical or is an aspect or attribute of it. And from such a mere negation of all positive qualities have 'emerged', we are asked to believe, our most intense desires, our clearest knowledge, our highest purposes, our aspirations towards the good, the true, the beautiful, our most effective efforts, our creative activities.

To this may be added a further comment that is applicable, not only to Lloyd Morgan and Strong and those who think with them, but to many others also. One main source of trouble both with philosophers and scientists is that they begin by treating Mind as a mere spectator of events; as Mr. Broad says, speaking of Whitehead and himself, 'We are concerned with . . . Mind only as something which perceives and thinks about matter.' If we begin by thinking of Mind as a mere spectator of events, it is difficult to give it any other status. The pitch is queered from the start of the game.

Mr. Edmund Noble's Version

Of all the schemes of Emergent Evolution that of Mr. Edmund Noble seems in many respects the most consistent and acceptable. His book,[1] as is implied by the title and

[1] *Purposive Evolution, the Link Between Science and Religion* (London, 1926).

many of the chapter-headings, professes to accept purposive or teleological causation as real and efficacious ; and he begins by rejecting very decidedly, not to say scornfully, all attempts to import something of the nature of Mind, of the psychical, into physical nature. Our keenest anticipations are therefore evoked ; but are destined to be grievously disappointed. For his solution of the problem (that of evolving purposive intelligent action from a mechanistic physical world) consists merely in ignoring the essential nature of teleological causation and redefining it to suit his purpose. He draws a sharp line between organic and inorganic causation, holding that synthesis, configuration, or system-determination, is characteristic of the former and absent in the latter. 'In inorganic causation we get a multitude of separate incidences, no one of which, as object, is immediately necessary to the rest. Organic causation, on the other hand, is the product of contributing numbers—a power cumulative from part to whole and refluent from system to unit.' The natural environment, from electron to star, is, he says, 'ruled by inorganic causation alone'. Then, after citing a number of instances of organic causation, he writes : 'In all these examples, then, we observe implications of inter-dependence and reciprocality which are utterly lacking in inorganic causation.' But 'the two forms of control (of causation) are fundamentally one'. For 'while organic causation depends for its origin and possibility on inorganic, there is no such dependence of inorganic causation on organic'. In short, organic causation, so defined, emerges from inorganic, and, having emerged, becomes the distinctive feature of organisms. 'In the non-living world there is no dependence of the structure of the unit on the structure of the aggregate. . . . Under inorganic causation, that is to say, the molecules have a merely inorganic relation to each other. . . . But when favoured by new conditions a new relation between them begins. What each of the molecules shall do is henceforth determined by what all of them must do ; the single properties of the units, in a word, are transmuted, or merged, into a collective property, the property of life. What we envisage, therefore, in life activities as their driving power, is not a "vital force" different from force in general, nor yet the physico-chemical properties of the molecules engaged, but just this power of the all-imposing collective character on each, just this superposition of organic property upon inorganic property, just this lifting of matter from simpler forms in which it is universe-maintained into complex forms which are self-maintained. The driving and directing power seen in life is then the organism as a whole.'

This scheme of emergent evolution differs, then, from others in making the principles of synthesis, of configuration, of emergence peculiar to the organic realm, denying them to the physical. It distinguishes itself also by the straightforward denial of all psychic nature or qualities to the inorganic or physical realm. And it identifies intelligent purposive action with a special complex form of mechanistic causation called organic causation, that is to say, it proposes to regard such organic causation as teleological, or at least, as that kind of causation to which alone the word teleological is properly applicable.

Conscious or psychical events are recognized as having emerged or as having been evolved from the purely physical realm (without psychical correlates) by way of protoplasmic events; but they are not of any particular importance, they do not modify the 'organic causation' of the bodily events in any important manner. The actual emergence of experience, cognitive and conative, is described in terms strongly reminiscent of Herbert Spencer. 'The commencing organism thus acquires its first experience of environment through unavoidable contact with objects, to which it applies the physical test of ease or difficulty for assimilation. As the ease of dealing with assimilable substances is actually a preference for them, so the difficulty of dealing with non-assimilable substances is tantamount to their rejection; *this ease and this difficulty, moreover, are gradually converted into the shocks and commotions which we call feelings.* The guidance, that is to say, is first that of mere physical ease and physical difficulty; through the development of awareness it finally becomes a guidance of sensations agreeable and disagreeable, such as, beginning in selection, finally culminates in activities of search and pursuit. Activities in relation to environment are involved throughout, and these may begin with the simplest kind of reaction from buffetings and distortions, due to the medium; once originated, they *gradually take on purposive character.*' [1] Again: 'Originally the shock which we call feeling could not have been long delayed once organic matter had begun

[1] P. 352. Italics mine.

to function for self-maintenance.' Thus easily may we account for the appearance of Mind in the universe, if only we allow ourselves to use language with utmost laxity and abstain from asking awkward questions.

Is then 'organic causation' in itself a pursuit of ends, a striving towards a goal? No; in addition to the emergence of 'organic causation' living organisms differ from inorganic things in a second way; namely, they 'are endowed with the function of self-maintenance'. In all cases of organic adaptation, 'given the impulse to end, to self-maintenance, the final stage is reached by the working of the parts engaged towards conditions of least resistance, and therefore towards the configurations which represent the maximum attainable economy and efficiency for the activities carried on'. Yes! Perhaps! *If* 'the *impulse to end*, to self-maintenance' is *given*, if the organism is *endowed* with this function. But how given, how endowed? It is Noble's business, the self-chosen task of this emergent evolutionist, to account for the appearance of intelligence and purpose in the world. He professes to give us 'a deeper view' which 'overcomes' the supposed distinction between teleology and mechanism. And this view, it would seem, is that purposiveness is to be equated with the principle of configuration.[1] 'In the organism also the action-reaction relations needed for the final grouping are the contribution of the individual molecules; the directing is the work of the total system. It is out of this "concilium" of units, each submitting itself to the rule of all, that purposive activities and forms take their rise and the system develops its means to self-maintenance.'

Purposiveness is, then, the expression of the tendency to self-maintenance and this in turn is an expression of the principle of configuration, of determination of the unit by the system.

But Noble is not content with this derivation of purpose from 'organic causation'. He has told us that the latter is peculiar to living organisms; he has denied it and the impulse of self-maintenance to inorganic systems. Yet he is determined to carry purposiveness down into the inorganic. 'Organic purposiveness widens out into the cosmic purposiveness from which it is derived; cosmic purposiveness

[1] In this he is at one with the psychologists of the *Gestalt* school; for their only recognition of purposiveness is the recognition of what they call the tendency to 'closure' in configurations.

reappears in the organic purposiveness which is its product and continuance. On the one hand we gain the view of a directing power not separate from things but identical with them—a power which, automatically eliminating friction and disturbance, sets up in stellar, planetary and material systems generally the order and harmony we call purposive.' He feels entitled to assert that the whole universe is purposive ; and he illustrates the purposiveness of inorganic events in the following way : 'The process which, by converting differential stresses into equalized stresses, results in teleological forms—forms which, inter-assimilated or self-assimilated, tend to endure—finds familiar illustration in the rounded shapes set up by friction ', as in the pebbles on the beach. The pebbles on the beach are then ' teleological forms ' because they are stable or enduring forms, and the pounding of the waves is purposive because it produces these ' teleological forms '. ' Cosmic self-maintenance, then, works out as purposiveness, and does this by setting up the enduring forms which, in virtue of both process and result, minus conscious design, suggest plan, seem as if shaped, aggregated or kept in balanced motion as the result of purpose. The end to be reached is endurance ; and it is reached by the automatic elimination of the differential stresses which cause motion and bring on change. . . . That man's subjection to his own society takes effect, not by a series of pushes and pulls, but through mental states involving consciousness, memory and will, neither makes the control less effective nor differentiates it in any fundamental way from the like process in the physical realm.'

For intelligent action is not to be identified with conscious action ; it is a much wider category and consciousness makes no difference ; ' intelligence . . . is not essentially psychical, but fundamentally dynamical.' And : ' The essence of Mind is not awareness, feeling, consciousness, but the organic modes in which power works towards enduring activities and forms. Does Mind act on body ? If we mean by Mind a " psychic power " which confers on body whatever intelligent character it may show, the answer is that it does not ; if we mean by Mind those fundamental modes which make power intelligent alike in nature [as in the breakers pounding the pebbles on the beach] and in life, the answer is that it does. And our answer must be the same when we ask whether consciousness has any causative or determining part in organic processes. If we mean by consciousness merely our awareness of what we are doing, the reply is that it has not; if we mean by consciousness the physical correlate of that awareness in brain and nervous system, the answer must come that it is so in-and-in-woven with the physical changes which it accompanies as to be an inseparable factor of them.'[1]

Teleological or purposive process is then, for Noble, merely process tending to produce enduring forms, collocations that are stable and therefore seem, as he says, to have been designed; awareness, foresight, desire, will and even 'organic causation' as defined by him, are not essential to it.[2]

Teleology in this sense pervades the physical realm and is merely an abstract aspect of such mechanistic events as result in greater stability, in more enduring equilibrium of systems, in the sphericity of pebbles or the accumulation of water in the hollows of the earth's surface. A mountain stream is as purposive as Alexander or Napoleon; for it is constantly producing 'teleological' results, namely, relatively stable forms.

Noble makes the mistake (made also by Broad and many others) of defining the teleological as a particular kind of structure or organization, namely, such as seems to have been designed and constructed in the service of some purpose. He never considers or inquires into actual purposive events or activities. It never seems to occur to him that such events are open to our inspection, that we are familiar with them, and that, if we had no such acquaintance, we should not conceive of any structure as being in any sense teleological. As Lloyd Morgan postulates Divine Purpose, so Noble postulates teleology (in his peculiarly mutilated sense of the word) throughout nature;

[1] The unmistakable confusion of this last sentence is surely significant of a confused state of mind resulting from a determined effort to achieve the impossible. (P. 452.)

[2] 'Teleology emerges, as we have seen, from a cosmos which is engaged in activities both separative and integrating, but supremely and finally in the collocation of its material units into relatively enduring systems by way of a motion process which works automatically to eliminate conditions of differential stress and to set up conditions of equalized stress. The end to be attained is the end of maintenance, and the reaching of it brings into existence the forms and configurations we call purposive—structures and systems which seem designed, the outcome of plan.'

and yet, even more completely than Lloyd Morgan, he ignores the types of experience through which alone we are led to conceive of anything as teleological. That is to say, like so many others, his treatment of the problem of teleology is utterly inadequate, because it does not set out by recognizing the psychological facts.

Noble commits also another error very commonly made in discussions of purposive action. He points to instances of self-consciously planned and foreseen action and then to other actions which seem no less teleological except that the actor is able to give no introspective account of the foreseeing or, perhaps, is able to deny any detailed foresight of the result produced. Hence, he argues, detailed foresight, since it plays no part in cases of the latter kind, when it is present must be regarded as otiose and non-operative. Against this oft-repeated and misleading argument it cannot be too often repeated that in purposive action our foresight is never adequate to the foreseen event ; the event itself is always richer and, in many cases, other in detail than our foresight of it. And, in some of the most unmistakably purposive and creative events, the foresight is only very vague ; but some foresight and the forward-looking urge or desire is there. I adduce again as a typical instance of such events the composition of the Marseillaise by Rouget de Lisle.[1]

I pass over the inconsistency involved in making purposiveness an expression of organic causation, peculiar

[1] It is related that he returned home from a gathering of like-minded friends who had urged him to compose a song that should express their patriotic fervour. He himself was consumed with zeal for the cause. His host had remarked ' that a poet must have inspiration and sent his daughter for the last bottle. They drank, and the young officer went back to his barracks through the cold night. His country's problems troubled him. What could a poor army officer do to correct them ? Scarcely more than sit down at his clavichord and pour out his emotions. . . . The inspiration came, note by note, and word by word the Marseillaise floated into the night. De Lisle never wrote a word of either until the next morning, when he awoke to find his mind still moved by the wild paean.' (G. P. Sousa in *New York Times*, August 26, 1928.) Even at so high a mental level, the purposive activity of the composer was but a groping towards a goal which, as also the road as means to it, was but vaguely foreseen. And all purposive action has something of this groping character. But to grope however blindly is to grope towards a goal, is to act teleologically. It is only where the goal we desire and the means for the attainment of it are of very familiar type that purposive action assumes its most fully developed form of devising and working out a plan perfected in detail for the attainment of a clearly and exactly envisaged goal.

to living things, and then extending it to the physical realm at large. I insist rather on the perversity which aims to destroy the distinction between mechanistic and teleological causation; and on the illegitimacy of the procedure by which it is sought to attain this goal. That procedure consists in stripping away some of the essential marks of purposive action and ignoring others, and in claiming as marks of the purposive what are essentially marks of the mechanistic.

Let us return to the rounded water-worn pebbles. It is true that a child or a savage finding a nearly spherical pebble regards it with curiosity or even awe. I have seen savages collect such pebbles and pile them upon their altars. But the man of culture has learnt to recognize such sphericity as a characteristic product of a multitude of mechanistic events. Contrast with this his attitude towards a chipped flint; consider the controversy over eoliths. What are the marks on which we properly rely in distinguishing from all others flints that owe their form to human purpose? Certainly not any approximation to the spherical form, but rather extreme departure from it in the direction of instability; a fine point and sharp edges, regularity of the flattened surfaces in size and arrangement, suitability as an instrument, as a means towards some common human goal, such as food or victory in combat!

It is the old story again: the teleological is opposed in direction to the mechanistic; it involves the building up, not of stable, but of unstable forms, the rearrangement of physical systems that have attained equilibrium in systems that are, as it were, poised in utmost defiance of the physical laws of equilibrium, systems which, as soon as they cease to be maintained by teleological activity, come crashing down in death and ruins. Can we imagine anything more unstable than a political system, the genius of an artist, a theory of the constitution of matter, or an aeroplane? But just these are the highest and most unmistakable instances of purposive systems, carrying to a high point that opposition to the products of mechanistic causation which is so well exemplified by the chipped flint of primitive man.

An animal may tear a stick to pieces with his teeth, and the purposiveness of such action remains obscure to those that have no eyes to see; but Professor Köhler's ape,

Sultan, gnawed away pieces from the end of one stick *in order to insert it* in the hollow end of another, *in order to have* a long stick, *in order to reach after*, and bring within his grasp, a banana. And the banana, when eaten, does not contribute to equilibrium; its substance is anabolized, built up into highly unstable molecules, part of taht store of energy in unstable equilibrium which enables the animal to be perpetually on the alert, perpetually seeking, exploring, endeavouring, striving for this, that, and the other natural goal.

This grandiose attempt by Noble to provide what he calls a deeper view of Nature, under which the distinction between the teleological and the mechanistic shall disappear, seems to be the most thoroughgoing of its kind and therefore deserves most careful consideration. I add, then, the following considerations. Suppose that our telescopes shall have been improved to the point that they magnify the features of the surface of Mars one hundred times more than any existing telescope. What features should we seek and accept as sure indications of the presence on Mars of intelligent purposive beings? Would it be forms in stable or in unstable equilibrium, enduring or perishable forms? If we descried a conical pile, its slopes at an angle of thirty degrees to the surface of the planet, or a hundred such piles, we should not accept them; for such would be in physical equilibrium. But, if we saw a single pile in the form of a slender tower with vertical walls, we should have little doubt; and, if we were to see many such, our doubts would vanish; for such a tower is in unstable equilibrium and, as such, presents an essential mark of the teleologically constructed. Or suppose that we saw, marked out in fire or water, a figure exactly like the figure illustrating the fifth or the forty-seventh proposition of the first book of Euclid. Such a figure is so unstable, so lacking in all physical equilibrium, that a single instance of it would convince all reasonable men that it was the product of design, the product not of mechanistic but of teleological causation. The chances, we should say, of its having been produced mechanistically are so infinitely small that its existence is proof of teleological constructive activity. Or take a simpler supposition. We see a river winding its way to sea across one of the land surfaces of Mars; and we discern a long section of it that runs in a perfectly straight line with parallel banks.

We know that such an arrangement is very unstable in physical nature: we should therefore strongly suspect teleological agency; and if we then found similar sections in other rivers, we should make this inference with confidence.

Noble's attempt to abolish the distinction between the mechanistic and the teleological consists, then, in denying and ignoring the essential marks of the teleological, and in setting up quite arbitrarily as the essential mark of the teleological the tendency to equilibrium found throughout the physical realm; or alternatively (for he vacillates and the two are not the same) the allied but very questionable appearance of order and harmony in physical nature.[1] While in respect to the chief self-chosen task of Emergent Evolution, the task of making plausible the emergence of Life and Mind from the physical realm, Noble's account is no whit superior to Herbert Spencer's: for he does not even press the principle of emergence into this service; he treats all psychical and conscious events as excrescences on the realm of nature too trivial to present a problem worthy of careful consideration.

[1] From Noble and many others we hear eloquent celebrations of the order and harmony of the physical realm. But it is just as easy to pick out and dilate upon the disorder and disharmony of that realm, to grow gloomily eloquent over the clash of worlds, the cataclysms, the destructive fury of the elements. Could there be anything more purely chaotic than the flaming envelope of our sun? Where is the order and harmony of a tornado or of any storm by sea or land, of a volcanic eruption, of an earthquake, a land-slide, a flood, a shipwreck, a railway collision, a conflagration, an explosion? The truth is that, when we talk of the order and harmony of physical nature, we are anthropomorphizing in the illegitimate or bad sense. Like the terms 'function' and 'efficiency', the words 'order', and 'harmony' are strictly inapplicable to the physical realm, unless we believe either that its events are intrinsically teleological or that they have been arranged or created for the realization of some purpose. If we accepted the suggestion of Noble that the purely physical realm is such as to tend to the production of order and harmony, we should, in view of the vast amount of conflict and disorder revealed by the history of mankind, be compelled to the conclusion that man's nature is such as to tend to disorder and disharmony. If we review the facts without prejudice, we see that there is no intrinsic tendency to order and harmony in the physical realm; and that man's effort, increasingly successful in proportion as civilization develops, is to prevent, subdue and overcome the disorders, the disharmonies, the catastrophes, of physical nature and of his own primary impulses; and, having attained some effective degree of self-direction, to establish in the physical world about him some measure of the regulated order and harmony that has been achieved within his own organism, directing blind physical forces to the service of his desires and his ideal aspirations.

It is remarkable that Mr. Noble should feel that, by reducing intelligence and purpose to mechanism and by calling mechanism intelligent and purposive (in the sense merely that mechanistic processes tend to the production of stable aggregations of matter) he is reconciling Science with Religion.

I add to the foregoing criticism of Noble's version of Emergent Evolution a few remarks on General Smuts' somewhat similar book *Holism*. The author exhibits in an interesting manner the many forms of aggregation presented by physical and organic nature, and arrives at the generalization that nature at all levels tends to the production of wholes ; and by ' whole ' he means much the same as other authors mean by syntheses, configurations or relatively closed systems. He then proposes the word ' Holism ' to express this general fact ; says that holism pervades all nature, and, going further, erects it into an active principle or agency that governs the course of all natural events. Holism thus becomes a governing agency which in the author's view renders unnecessary any other explanatory principles to account for the peculiarities of organisms.[1]

Alexander's Version

The version of Emergent Evolution presented by Dr. S. Alexander, in his Gifford Lectures,[2] is the most complete and impressive of all. I am well-nigh inhibited from any attempt to criticize it both by the awe I feel in face of so magnificent an effort and by my warm personal regard for the author. Yet Science must be served and I take my courage in both hands. The story begins with bare Space-Time as the matrix from which all things emerge, and culminates with the emergence of Deity. First to emerge is motion. Motions become complicated, and matter or some forerunner of matter emerges, and then the secondary qualities of matter, colours, sounds, odours, warmth and cold. Later comes Life and still later Mind, which in its higher developments becomes moral personality verging towards Deity.

I say nothing of the difficulty of conceiving motion with-

[1] Cp. a more detailed criticism in my article, ' The Confusion of the Concept ', *Journal of Philosophical Studies*, 1928.

[2] *Space, Time, and Deity* (London, 1920).

out anything that moves. And, though I cannot accept the neo-realism that ascribes to material things the qualities that all philosophers and psychologists, except the small band of neo-realists, have always regarded as qualities of sense-experience, I do not insist on the difficulty of this feature; for other men besides Alexander for whose intellects I have great respect find this strange view intelligible and acceptable.[1] I merely point out that neo-realism consists in taking away from Mind much that is commonly attributed to it and assigning it to matter; thus diminishing the interval between them.

More difficult still is the doctrine that knowing or cognition is a function not peculiar to Mind but one common to all material things. 'Thus the relation of the Mind to its object on the table is precisely of the same order as that between the floor and the table . . . knowing is nothing but the empirical form which compresence assumes when one of the partners has the empirical quality of consciousness.' . . . 'Cognition, then, instead of being a unique relation, is *nothing but*[2] an instance of the simplest and most universal of all relations.' Knowing is merely a special form of compresence or of being together. I take it that compresence involves causal relations between the objects said to be compresent. If so, the doctrine is similar to Lloyd Morgan's view that 'reference' is the universal concomitant of physical causation. But if we accept even this strange identification of knowing with compresence, the identification is not complete. Though cognition is said to be nothing but an instance of compresence or causality, it is an instance of a peculiar kind; it has 'the empirical quality of consciousness', for 'the mind in virtue of its conscious quality is aware or conscious of that object', the object with which it is compresent. So that, by Alexander's own admission, it is not true that, as he says, 'there is nothing in the compresence between the mind and its objects to distinguish that relation from the compresence between any two objects which it contemplates, like the tree and the grass'. There is in the former instance of compresence, a not insignificant distinguishing

[1] Even pain and pleasure are made qualities of material objects. Pain is a quality, not of the hammer that bruises my finger, but of the tissue that is bruised; and what we call the pleasure of eating (or of playing a game, or of doing a kindness) is a quality, not of the sugar we eat, but of the tissues involved in the processes.

[2] Italics mine.

quality, ' the empirical quality of consciousness ' ; knowing is not *nothing but* an instance, of compresence, it is compresence plus *awareness* or *consciousness of*. We have therefore the whole problem of the emergence of *consciousnesss of* from merely compresent or reciprocally interacting movements in Space-Time. And this problem (to which, as we have seen, Lloyd Morgan devotes his utmost efforts without avail) is merely passed over by Alexander. Further, the organism does not react cognitively to all objects with which it is compresent ; it selects among them, as Alexander himself insists. And it is not that, like a physical object, it is unaffected or but slightly affected by certain of its compresents. It selects among the many things which are simultaneously acting causally upon it, cognizes the one to the neglect of the others, and then, perhaps, cognizes several of those others in turn. Such selectivity of Mind is without parallel in the physical realm. It is conative. Let us, then, see what Alexander has to say of conation.

Mind, having been stripped of sense-qualities, of imagery,[1] and of pleasure and pain, remains a series of enjoyed conative acts. Conation is the causal activity of minds.[2] ' The mental act is a conation, which is something mental, and not merely physiological. . . . Primarily, conation is practical, and it issues in movements which tend to alter or destroy the object or at least to affect our relation to the object.' Alexander, then, does not deny the action of Mind on body ; he asserts it.[3] He does not, like Lloyd Morgan, perpetually take away with one hand what he gives us with the other; he does not, like Strong, assert that the physical is psychical ; he does not, like Noble, scornfully reject, as the expression of a primitive and pathetic ' psychomorphic ' fallacy, the belief that Mind in some degree controls the body and, through it, the course of events in the physical realm ; rather, he frankly admits and asserts it.[4] But we are told strangely little of

[1] For imagery also is identified with physical things.

[2] He explicitly rejects Epiphenomenalism and Psycho-physical Parallelism and writes : ' On our view it still remains true that Mind and brain interact if the phrase is properly interpreted.'

[3] One might say that, in this system, conation is ' the last infirmity of noble mind ', the last rag of clothing that serves to distinguish it from naked physical nature.

[4] ' Hence, although Mind cannot be and act without things from which to select its objects, neither the things nor the objects are affected in themselves by the presence of Mind except so far as the

the process or activity which is conation. The words 'purpose' and 'teleology' do not occur in the index; and in the whole of this comprehensive cosmogony there is no discussion of the relation of teleological to mechanical causation. There is an excellently sane discussion of causation. It is neither denied nor explained away. 'A cause is the motion of a substance, or a substance in respect of its motion. . . . Causality is thus the spatio-temporal continuity of one substance with another; and the cause is the motion which precedes that into which, let us say, it passes or is transformed.' All causation then is communication of motion, is causation in the most strictly mechanical sense.

It is true that a few passages are difficult to reconcile with this view of all causation as the communication of motion.[1] The future is said to be enjoyed. But 'Such enjoyment is the future in idea, and this is the only way in which the future as future can be enjoyed. This future enjoyment is causal to its own realization as a present. But this enjoyment drives us not *a fronte* but *a tergo* like all other causality.' It seems clear then that for Alexander teleological causation is only a special case of mechanistic causation and is no exception to his generalization that all causation is the transition of motion in Space-Time. His account of mental activity is that it is merely an incident in a chain of mechanistic causation; and is none other than the old associationist account—the idea of a future movement causes the movement.

All the evidence that purposive action is (objectively regarded) quite peculiar and distinctive and that the growth and activities of organisms are teleological and cannot be mechanistically explained, all this is ignored. Unlike Hobhouse, Alexander does not first struggle to show how teleological causation can emerge from mechanical, and then give up the attempt; unlike Lloyd Morgan, he does not recognize that 'foresight' with enjoyment affords all-important guidance to action and then strive vainly to describe the emergence of such foresight from mechanical sequences; unlike Noble, he

mental conation alters them.' And 'our mind and external things are, as compresent existences, in causal relation to one another'.

[1] E.g. 'When a thought brings about its own realization in an act of will the immanent process is the transition of the thought into a perception and that is purely mental.' And there is one mention of teleological action.

does not profess to find in physical nature an all-pervading purpose (by dint of stripping the word 'purpose' of all its significance and meaning); rather, like Strong, he simply ignores the problem of the relation of teleology to mechanism, that problem which is the central and, it would seem, insoluble problem for any scheme of Emergent Evolution that professes to describe the emergence of Mind from the physical realm.

No one of our exponents of Emergent Evolution makes any approach to a solution of this, its most urgent problem. Strong and Alexander ignore it; Noble reaches a fictitious solution by importing a falsified and utterly castrated teleology into the physical realm; Lloyd Morgan struggles heroically with it, but achieves a tangle of inconsistencies; Hobhouse attacks the problem frankly, with full acknowledgment of the radical difference between the two kinds of causation, and in the end acknowledges that Mind works teleologically and matter mechanistically, and that living organisms are the seat of interplay of the two forms of causation, neither of which can be derived from the other as an emergent. To which it may be added that other emergentists, like Broad, concern themselves only with systems that seem to be teleological in the sense that, like machines, they seem to be designed; and they never come within sight of the problem of teleological causation.

Professor Sellars' Version

The version of Emergent Evolution told by Professor R. Y. Sellars in his *Evolutionary Naturalism* [1] is in my judgment the most thoroughgoing and consistent and the least objectionable of all. At the date of publication of this book the expression Emergent Evolution had hardly come into vogue and Sellars does not make use of it. But the principle of emergence is clearly stated under the head of creative synthesis or the appearance of true novelties as the organization of physical events becomes more complex, the latest and highest of such novelties being psychical events and conscious purposive thinking that has causal efficacy in guiding the cerebral processes of the human organism. There is no importation of the mental into the unevolved physical; Lloyd Morgan's Spinozistic universal concomitance of the physical and the psychical

[1] Chicago, 1922.

and Strong's pan-psychism are equally and decisively rejected, as also Epiphenomenalism, Psycho-physical Parallelism and Neo-realism. The psychical is evolved from a purely physical and mechanical world that has nothing of the nature of Mind; and the psychical, when it attains the human level with the high complexity of human brain-processes, plays its part as conscious thinking, as awareness, as planning and contriving, in the total complex of psycho-physical events.

Here at last, then, is one author who does not scruple to evolve Mind from the purely physical realm and to give it a role to play in natural events. The weakness of Sellars' presentation of the case for Emergent Evolution is that he never comes to close quarters with the alleged processes of emergence of Life and of Mind. He presents his critical realism convincingly and assures us that anyone who will combine with it what is sound in pragmatism will naturally accept his evolutionary naturalism. Now I for one have followed that prescription; but do not find myself able to accept Sellars' evolutionary naturalism. That doctrine does not seem to emerge inevitably or even plausibly from the synthesis. It is, I think, just by reason of his neglect to grapple closely with these problems of emergence that Sellars is able to accept his own scheme so confidently and light-heartedly. All the other exponents of Emergent Evolution seem to agree with me in this matter. Sellars' scheme was early in the field and must be known to them, yet they do not accept it. Why do they all balk at the crucial transformation? Like other conjurers, they know that if you are to produce Mind from a hat (or from any other physical arrangement) you must first put it there or have it up your sleeve; or else you must be content to produce a mere semblance of Mind; and they take the necessary precautions. But Sellars merely exhibits his set-up and, without attempting to perform the operation or to show in detail how it may be done, says: Now, you see, it is perfectly reasonable to suppose that Mind will emerge. It is true that he softens the emergence of cognition by allowing the prior emergence of psychical events that are not awareness of anything; and he softens the emergence of purposive striving by asserting that the emergence of events that are purposive is preceded by the emergence of events that are no longer mechanical or mechanistic. In both cases he is postulating events of a kind for which we have no warrant. Further, Sellars does not grapple in any way with the facts of heredity and morphogenesis. These are events that occur below the level of his emergence of cognition and purpose, the level of highly complex brains, yet they have the marks of being in some lowly sense teleological. Nor does he deal in any way with the so urgent problem of memory.

NOTE 13. *Dr. Lloyd Morgan on Consonance of Welfare and Pleasure* [1]

Many years ago I argued that the high degree of positive correlation that seems to obtain between, on the one hand, pleasantness and unpleasantness of experiences and, on the other hand, the beneficial and the harmful nature of experiences, respectively, sets a dilemma before us; namely, we must believe either in the causal efficacy of pleasure and pain in the complex of psycho-physical events that issue in behaviour; or we must believe that the positive correlation of pleasure with the beneficial and of pain with the hurtful was established at the outset of organic evolution (or at some point in its course) by a beneficent Providence.[2]

The argument was that, if pleasure really sustains action and tends to bring about repetition of similar action under similar circumstances, as it seems to do, then we can understand that those creatures which found pleasure in actions that are beneficial to themselves will have had in the struggle for existence great advantage over those to which such actions were not pleasurable; for they will have tended to repeat such beneficial actions. Still more will they have had advantage over those of their species to whom such beneficial actions brought pain; for these would have avoided all repetition of such actions. In this way, natural selection would have produced a race descended in the main from individuals in which beneficial actions were pleasurable; and if this peculiarity were inherited, the race would exhibit the positive correlation that actually obtains.

Similar considerations would account for the positive correlation between the painful and the harmful. Those members of a species that varied in the direction of finding pleasure in harmful modes of behaviour would have been eliminated in the struggle for existence; those that varied in the opposite direction (finding harmful modes of behaviour painful) would in the main have had advantage

[1] I have to thank the Editor of *Mind* for permission to reprint this note from the pages of his journal.

[2] *Primer of Physiological Psychology* (London, 1905).

in the struggle for existence because they would have avoided repetition of hurtful modes of behaviour.

In his second volume of Gifford Lectures,[1] Dr. Lloyd Morgan has taken up this challenge to all who deny psycho-physical interaction and professes to refute my argument. Describing the correlation from which my argument sets out as 'Consonance of Welfare and Pleasure', he begins by admitting the general consonance or correlation alleged and asks: 'Is this interpretable under concomitance? I maintain that it is so interpretable.' The word 'concomitance' is here used to imply the Spinozistic view that denies psycho-physical interaction; but it may stand for all theories that deny causal efficacy of mental or psychical events in psycho-physical process. The reasoning by which Lloyd Morgan claims to have refuted my argument is wholly contained in the following passage: 'That such consonance is a principle of universal validity, as Mr. McDougall here assumes,[2] is contrary to many patent facts of human life, and is unproven in animal life, as he himself insists in the *Outline*. In many recorded instances animals removed to a new habitat seem to have pleasurable enjoyment in eating plants that are poisonous and lead to sickness, or even to death. Such instances serve to show that consonance has been established only in relation to the normal environment within which the animal has been evolved. It is widely prevalent under cognitive guidance; but it is a derived and not an original alliance. . . . Conscious guidance is nowise infallible; it may often lead to elimination rather than survival. This, however, entails elimination of those organisms in which there is lack of consonance. Hence, in so far as consonance obtains, it counts for life-progress, and this every whit as much on our nterpretation as on that of Mr. McDougall.' That is to say, Lloyd Morgan agrees with me in holding that the consonance which actually obtains must be regarded as having been established by natural selection. In fact, and it is a disturbing fact, the argument by which he claims to have refuted my argument is in essence a restatement of my argument. I say it is a disturbing fact that two persons, accepting and honestly considering the same facts,

[1] *Life, Mind, and Spirit* (London, 1926).

[2] Of course I do not and did not make any such assumption. The acts were as patent to me as they are to Dr. Lloyd Morgan or to any other intelligent person.

should disagree so completely as to their bearing on theory. It is a fact that tends to 'scepticism of the instrument'.

This is only one point of difference among many, but it is one in which all our differences centre; it is a crucial question. Unless the human mind is utterly unfitted to deal with such questions, it must surely be possible to reach a generally acceptable decision in favour of Lloyd Morgan's view or of mine.

Lloyd Morgan's escape from the dilemma I proposed consists in boldly grasping one horn and asserting that he remains unscathed. The dilemma proposed was: either natural selection establishing a certain correlation between modes of experience and modes of behaviour or beneficent Providence intervening in the course of evolution. Lloyd Morgan replies: Yes, the general consonance of welfare and pleasure must certainly be regarded as an effect of natural selection, a consequence of the elimination of creatures that varied in the opposite direction and found pleasure in harmful modes of behaviour; or, concisely, consonance is general because it effectively contributes to survival. So far then we are agreed. But my further point is that such consonance can have contributed to survival only if pleasure and pain really have the causal efficacy in behaviour which they seem to have. Behaviour belongs to Lloyd Morgan's life-story, pleasure to his mind-story; and, as he tells us again and again, the two stories are concomitant, but the events of the one are without influence on those of the other. How then should events of a certain kind in the one story (pleasant experiences) become correlated with events of a certain kind in the other through natural selection?

My argument was stated in abstract terms. It seems necessary to restate it as concretely as possible. Imagine, then, a group of animals of some herbivorous species, say deer, entering and making their habitat a region in which, for the first time in the history of that species, noxious bitter-tasting herbs abound. At first all the deer are indifferent to the bitter taste of the noxious herb. The noxious substance in the leaves of the herb stimulates the taste-nerves of them all, and all experience bitterness; but bitterness has for them no marked hedonic tone, is neither markedly pleasant nor unpleasant; and all eat the herb as they graze.

The story must be continued in two versions, Lloyd

Morgan's version, according to the principle of concomitance without psycho-physical interaction, and my version. First, then, the version according to the principle of interaction, my version.

Like ourselves the deer of later generations vary in respect to the effect of the bitter substance upon them ; all experience bitterness, but for some the bitter state is not unpleasant, perhaps for some (as for some of us) in its milder intensities it is pleasant. For others the bitterness in its faintest intensity is unpleasant. The former continue to eat it, for it is not repugnant to them. The latter turn aside as soon as they begin to nibble a leaf of the herb. In each generation many of the former succumb to the noxious herb ; the latter survive. The new tendency produced by the variation is transmitted to their progeny and soon becomes a general innate tendency of the group. Natural selection has produced consonance of welfare and pleasure, or, rather, between ill-fare and displeasure.

Note that the variation postulated in the foregoing version is a psychic variation, a variation in the direction of association of bitter sense-quality with unpleasantness. We know from human experience that we humans do differ from one another in this respect ; for some of us the bitter taste-quality, at least in its lower intensities, is pleasant ; for others of us the least trace of bitterness is unpleasant. And we behave accordingly ; the former seek and partake freely of bitter substances ; the latter reject and avoid all such. It is then reasonable to suppose that there may be similar differences among animals and that they may represent hereditary variations.

Now the version according to the principle of concomitance : Lloyd Morgan's version. Similar psychic variations occur in the group of deer and their progeny ; all begin to eat the bitter herb ; and for some the bitter taste-quality is unpleasant ; for others it is neutral or pleasant. According to the hypothesis of concomitance, the unpleasantness makes no difference to bodily processes, does not affect behaviour. Therefore, all the deer continue to eat the herb and all suffer its noxious effects in equal degree. Those in which the bitter taste-quality is unpleasant have thereby no advantage in the struggle for existence ; hence natural selection can produce no consonance between pleasure and welfare, or between displeasure and ill-fare.

Notice that I do not maintain that natural selection could not, according to the concomitance hypothesis, produce a race that avoided the noxious herb. Obviously it might do so, if the herb had some character, such, for example, as a peculiar colour, say blue, that might affect the visional sense in a differential or peculiar manner. Then, under concomitance, there might occur variations such that blue leaves would be avoided; and those deer in which occurred this variation of positive survival value would have great advantage—the group would become one in which all herbs of this colour are avoided. But it would not follow that in these survivors the bitter-taste should be unpleasant; the taste-quality might remain of neutral hedonic tone or might vary in either direction; and since natural selection works only through bodily processes and behaviour it could have nothing to say to the hedonic tone. No consonance would result.

Of course in any particular instance there might occur consonance of pleasure and welfare; but in the main dissonance would, so far as natural selection is concerned, be just as common. Yet Lloyd Morgan accepts general consonance as a fact, as a positive correlation produced by natural selection.

I know well that some of those who deny psycho-physical interaction will say that my argument is still stated in terms too general and abstract. Let us get down to the intimate physiological events concerned, they will say, and then the fallacy of McDougall's consonance argument will clearly appear. Now there is no generally accepted or acceptable theory of the neural concomitants of pleasure and pain. I select, therefore, one theory of such concomitance that seems particularly favourable to Lloyd Morgan's contentions, namely, the theory put forward by one who is physicist, physiologist and psychologist, Dr. L. T. Troland.[1] Troland's theory is that pleasure is the psychic concomitant of a general predominance in the brain of falling synaptic resistances, and that pain is similarly the psychic concomitant of general predominance in the brain of rising synaptic resistances.[2] Let us call

[1] In *The Mystery of Mind* and various earlier publications.

[2] I might point out in passing that I was the first to insist on the importance in brain-events of synaptic resistances varying with the general functional conditions in the brain (Articles in *Mind*, N.S., vol. VII). I am, therefore, well prepared to admit them.

these two brain-processes f. s. r. and r. s. r. respectively. And let us make a third version of the deer story in these terms. Now it is a feature and a perfectly reasonable feature of Troland's hypothesis that f. s. r. means continuance of the activities during which it occurs and a tendency to repeat them; that r. s. r. means the opposite, namely, discontinuance of the activities and less tendency to repeat them than if f. s. r. occurs. It is obvious that natural selection may, on these assumptions, produce in the race of deer a positive correlation of r. s. r. with the eating of the noxious herb. Those that vary in such a way that the stimulus to the taste-nerves results in r. s. r. will have advantage; the result of natural selection through the generations will be that in the race the specific stimulation by the herb will lead to cessation of eating the herb and avoidance of it. Then pain or unpleasantness, being the psychic concomitant of r. s. r., will also have become correlated in the race with this kind of stimulation and its effects in the brain. What more need be said? Lloyd Morgan's case is made out.

But we must go farther back. My question essentially is: Who or what determined that pleasure should be the psychic concomitant of f. s. r. and pain that of r. s. r.? This is the correlation that demands explanation. Suppose the opposite psycho-physical correlation, or concomitance, namely, pleasure with r. s. r. and pain with f. s. r. Then natural selection would produce a race of deer which would reject and avoid the herb although its taste was pleasant to them; a race in which there would be dissonance of welfare and pleasure, in which the pleasant would be concomitant with harmful and pain with beneficial modes of behaviour.

In my original presentation of the argument I pointed out that, if pleasure and pain differed only as two sense-qualities differ, as red differs from blue, or salt from sour, or warmth from coolness, or one note of the musical scale from another, then my argument would have no *locus standi* and no punch. It would in short deserve the neglect that it has 'enjoyed' and the cavalier treatment given it by Lloyd Morgan after the lapse of nearly a quarter of a century. But pleasure and pain do not differ in that way. There is nothing unintelligible or irrational or absurd in the fact that some of us like and seek bitter tastes and some dislike and avoid them, or in the fact that some of us find purples pleasant and others find them displeasing. But if we found some race of remote islanders who persistently sought and preferred those things and repeated those modes of behaviour that they honestly asserted to be unpleasant and avoided consistently all things pleasant, a race in short in which there was dissonance of pleasure and welfare rather than consonance, a race (in terms of Troland's theory) in which pleasure was the psychic concomitant of r. s. r. and pain that of f. s. r, then we should judge, and rightly, that there was something profoundly at fault, something irrational and absurd in their constitution. But under denial of psycho-physical interaction such races are just as likely to evolve under natural selection as the races we actually find, namely, races in which, in general and in the main, consonance obtains. I cannot put the matter more clearly. Dr. Troland has been my cordial colleague for seven years, yet I have never succeeded in giving my argument the least purchase upon his mind. Can I hope that Dr. Lloyd Morgan will be more open to it ? It would seem that there is some radical discrepancy between the mode of working of his mind and mine, a discrepancy which is brought home to me forcibly in every chapter of his two volumes of Gifford 'Lectures. In both volumes he repeatedly insists that he accepts universal concomitance of the psychical and the physical without interaction or reciprocal influence of any kind. He makes this above doubt clear again and again. Yet in equally numerous passages in both volumes he writes of conscious guidance of behaviour ; he distinguishes between the behaviour of lower organisms where, he says, there is no evidence of prospective reference and none of conscious guidance and that of higher animals and men, where, 'As I read the evidence, in *some* behaviour there is guidance under prospective reference.' He writes : 'If there be what I call foretaste in enjoyment coupled with cognitive prevision of coming events, we have an affective factor in guidance of no little importance—nay, more, as I think, of the greatest importance.' Again : 'I believe that conscious guidance does count for progress [i.e., in evolution], and that *au fond* all conscious guidance at the cognitive level is towards pleasure and away from pain or 'discomfort.' . . . There is cognitive guidance under affective signature.' And writing of a lowly animal he attributes to it 'cognitive reference which, as prospective, subserves some small manner of conscious guidance of events which we observe.' . . . The observed events being bodily movements of the animal. Once more : 'My interpretation of the chick's status is frankly behaviouristic, if a correlated psychical system not yet effective in guidance be acknowledged. But *pari passu* with the evolution of its behaviour there is developed proficient reference to that towards which it behaves. And with this comes conscious guidance which the behaviourist on his part will not allow.' In this last passage we have him drawing clearly a distinction between a concomitant psychical system 'not yet effective in guidance', and more highly evolved psychical systems that are effective in guidance of behaviour.

In the Gifford Lectures Lloyd Morgan nowhere admits the causal efficacy of human purposive striving towards a goal, yet in his essay in the recent volume, *Creation by Evolution*, we read not only of conscious guidance of behaviour and of conduct, but also of 'the dawn of that freedom of choice which we cherish above all things . . . the very turning-point in the evolutionary history of events. In that history it is of all events the greatest in promise. In human life it marks us as what we verily are: makers of a new and, as we hope, a better world. For human guidance is always toward something more or less clearly envisaged as not yet in being, but still to be brought into being through striving and endeavour.' Yet this is the same writer who in his Gifford Lectures rejects my hormic psychology root and branch; brushes it aside as merely literary rather than scientific, just because it recognizes the reality and efficiency of such striving, and because it endeavours to bring human freedom and purpose into evolutionary line with the lower forms of Mind.

This question of the origin of the consonance of pleasure and welfare is only a special case of a larger, more general problem which is acutely raised by Lloyd Morgan's combination of Emergent Evolution with universal concomitance; namely, if there is no psycho-physical interaction, and if our minds like their physical concomitants are the product of Emergent Evolution, how comes it that the mind of each of us is in some sense a microcosm, how explain the fact that in some sense, and with some degree of faithfulness, it reflects the nature and relations of the physical world? For, though correspondence is not the test of truth, there must be correspondence of some degree between the nature of the physical world and our knowledge of it. It may not be strictly true that the physical world exists in or presents tri-dimensional spatial relations such as I conceive them ; but Lloyd Morgan's naturalistic philosophy assumes the reality of physical objects in a system of relations corresponding pretty closely to the spatial relations we conceive. How, then, has this correspondence between physical space and our consciousness of it been achieved in the course of evolution ? As with the consonance of welfare and pleasure, only two answers are possible : either psycho-physical interaction or Divine intervention all along the line of evolution. Perhaps the latter alternative is accepted by Lloyd Morgan ; for in the Gifford Lectures we hear much of Causality, the directive Activity of God. 'For better or worse, I acknowledge God as the nisus through whose Activity emergents emerge, and the whole course of emergent evolution is directed.' If such preservation of correspondence between emerging mind and the physical world by Divine directive Activity is Lloyd Morgan's solution of the problem of knowledge he has not stated the fact in language intelligible to me, although he has discussed the problem of knowledge at some length. And, if it is his solution of the problem of knowledge, why does he not accept the same easy solution of the problem of consonance of welfare and pleasure ?

NOTE 14. *The Recognition of Teleology by some Mechanistic Biologists.*

It is significant that, although the majority of biologists are thoroughgoing mechanists, the more thoughtful of them make some kind of qualification of their creed. The common practice is, while asserting that science must seek and find only mechanistic explanations, to admit that metaphysics may properly offer teleological explanations that have a certain justification and validity. Thus Claude Bernard : 'Life is the directive idea or evolutive force of the creature ; but it would be an error to believe that this metaphysical force is active after the manner of a physical force. . . . The metaphysical evolutive force by which we may characterize life is useless to science, because, existing apart from physical forces, it can exercise no influence upon them. . . . If we can define life with the help of a special metaphysical conception, it is none the less true that mechanical, physical, and chemical forces are the sole effective agents of the living organism, and that the physiologist has to take account of their action alone.'[1]

This seems to me to involve a vicious dualism, the assumption that there are two kinds of truth about nature, two different and incompatible accounts, the scientific and the metaphysical, both of which may be true and exhaustive. But metaphysics has no sources of knowledge that are denied to science ; and the most necessary of all principles at the foundation of science is that of two contradictory statements, one must be false, if the other is true. If then it is true that there is only one kind of causation, the mechanistic, throughout the realm of nature, it cannot also be true that causation of a radically different kind, namely, teleological, also occurs. And, if teleological causation occurs, science cannot leave it to metaphysics and content itself with seeking only mechanistic explanations.

Among contemporary men of science two biological chemists stand out as exponents of this dualism of science and metaphysics, namely, Dr. L. J. Henderson (in his *Fitness of the Environment*, N.Y., 1913) and Mr. J. Needham (in *Man a Machine*, London, and N.Y., 1928).

[1] I cite from *A History of European Thought*, by Th. Merz.

Henderson shows that in many different ways the chemical properties of the earth's crust, and especially the properties of water and of carbonic acid, are just such as are necessary for the existence of living things, just such that, if they were different in any one of a number of respects, the life of organisms would be impossible. This assemblage of properties, all of which are essential to life, justifies us, he says, in speaking of the fitness of the environment, its fitness to be the abode of living things. This fitness of the environment cannot be denied; and the recognition of it undoubtedly raises a problem. 'The properties of matter and the course of cosmic evolution are now seen to be intimately related to the structure of the living being and to its activities; they become, therefore, far more important in biology than has been previously suspected. For the whole evolutionary process, both cosmic and organic, is one, and the biologist may now rightly regard the universe in its very essence as biocentric.' These are the closing sentences of Dr. Henderson's book. What does he mean by his conclusion that the universe is biocentric?

The old view that the material universe was made for the sake of man, that it was in some sense designed and made or arranged in order that mankind might live and flourish upon it, that its constitution was designedly adapted with the purpose of fitting it to be the abode of man, that view has commonly been called anthropocentric. Henderson's words can only mean the acceptance of that view with the modification that living things in general are put beside mankind, as part of the realm of life for the sake of which the constitution of the universe has been teleologically created, arranged, or moulded. This interpretation of his words is borne out by other passages, for example: 'Our new teleology cannot have originated in or through mechanism, but it is a necessary and pre-established associate of mechanism.' Again: 'If life has originated by an evolutionary process from dead matter, that is surely the crowning and most wonderful instance of teleology in the whole universe.' Again: 'We appear to be led to the assumption that the genetic or evolutionary processes, both cosmic and biological, when considered in certain aspects, constitute a single orderly development that yields results not merely contingent, but resembling those which in human action we recognize as purposeful. For, undeniably, two things which are related together in a complex manner by reciprocal fitness make up in a very real sense a unit—something quite different from the two alone, or the sum of the two, or the relationship between the two. In human affairs such a unit arises only from the effective operation of purpose.'

Henderson, then, accepts teleological causation, as having played an essential part in the production of living organisms. Yet he writes: 'At length we have reached the conclusion which I was concerned to establish. Science has finally put the old teleology to death. . . . The man of science is not even obliged to have an opinion concerning its reality, for it dwells in another world where he as scientist can never enter.' It seems clear, then, that Henderson accepts, as the inevitable conclusion of his careful argument from a vast range of well-established facts, the view that some teleological agent, some designing and creative Mind, in short God, made or arranged the material universe in such a way that, though the seat of mechanistic processes only, it has produced living organisms (including man) which also are the seats of purely mechanistic processes only. Henderson speaks of this view as 'our new teleology'; but in what sense it is new I fail to understand. Is it not identical with the view of the Deists of the eighteenth century? The teleology which Henderson rejects under the title 'the old teleology' is the teleology implied by the argument of this book, the view that Mind, wherever and whenever it exists, operates teleologically and is thus operative perhaps in all living organisms, certainly in man and probably in all animals. Now, as we have seen, Henderson regards his argument as proving that Mind has operated teleologically upon the mechanistic world of matter and physical energy. He believes, then, that Mind is capable of so operating. Why does he deny all such operation to Mind as manifested in men and animals? He gives us no reason for his denial.

We find, I think, the implicit ground of Henderson's denial stated in Needham's book. Needham's position towards teleology seems to be not quite the same as Henderson's. He quotes with approval Henderson's dictum that 'teleology dwells in another world which the scientific man as such can never enter'. He accepts Henderson's conclusion that there is as much teleology about the inorganic as about the organic'. He writes: 'What Henderson has demonstrated is that there is purposiveness in the inorganic universe no less than in living organisms. . . . Purposiveness, then, exists everywhere, it permeates the whole universe, and is not a unique characteristic of life. Sig. Rignano's teleology melts away in the immensity of that discussed by Lawrence Henderson . . . every individual piece of purposiveness bears, like a coin, stamped upon it the image and superscription of universal teleology.'

So far, it might seem that Needham's neo-mechanism (as he calls it) is, like Henderson's, a relegation of teleology to an original creative or ordering activity by which the Great Machine was set upon the path that was to produce man and all his works by purely mechanistic processes. But Needham introduces a complication. He confesses that the 'subjectivation of mechanism and teleology' greatly appeals to him; and chooses, therefore, to assert 'it is a mistake to suppose that there is anything really corresponding to either teleology or mechanism in external nature'. Holding, then, that both mechanistic and teleological explanation are purely subjective and illusory, that in nature there is neither mechanistic nor teleological causation, he holds himself free to choose either kind of explanation for scientific purposes. And he chooses the mechanistic explanations. He 'recognizes the supreme jurisdiction of the mechanistic theory of life, but admits it at the same time to be a methodological fiction'.

I shall not delay to insist on the unsatisfactory character of this subjectivist attitude, to point out that it is essentially solipsistic, that its denial of the reality of causation undermines all science. Let us rather examine the grounds offered by Needham for his ban upon all teleological explanation in Science. Holding himself free to choose between mechanistic and teleological principles of explanation, or to make use of both where the facts may seem to demand them, why does Needham choose to use the mechanistic only, why does he refuse to admit the propriety of teleological explanations in Science? His answer to this question is threefold. First: 'Mechanism and materialism lie at the foundation of scientific thought; without materialism science becomes impossible or at least unduly diminished in value. . . . Science, with the determinism and mechanism which form so unalterable a part of its existence, cannot be expected to take into account notions incompatible with its peculiar mental twists. Teleology is clearly one of these notions.' Now this reason, so naïvely stated, is of little weight. It is true that the science of the nineteenth century was founded upon Atomic Materialism and thoroughgoing mechanism; and it is true that many men of science brought up in its atmosphere have acquired a corresponding mental twist which prompts them to desire mechanistic explanations and closes their minds to all others. But though such a mental twist is not easily corrected in any individual who has once acquired it, it may be hoped that in the coming generations of men of science the acquisition of the twist may be avoided. Needham's first reason for his choice is no reason; it is rather a teleological explanation of the mechanistic twist that he, like so many other laboratory workers, has acquired.

Secondly, Henderson has shown that teleological causation must have shaped and ordered the constitution of the material universe in order to make it a fit environment for living things. If then the Creator has designed and produced once for all the fitness of the environment, He may also have achieved the far more difficult task of arranging for the fitness of organisms to that environment, that is to say, he may have so designed the Great Machine that its purely mechanistic events shall produce man and all his works. This is the only possible interpretation of Needham's assertion that the teleology manifested by organisms 'melts away in the immensity of that discussed by Lawrence Henderson'. To this we reply—It may be so; but the conclusion does not necessarily follow from Henderson's demonstration. Rather it remains vastly improbable. To design the properties of water and of carbonic acid in such wise that they shall be well fitted for the environmental and the internal events of living things is one thing; to design the Great Machine in such a way that it shall produce the plays of Shakespeare is another and infinitely more difficult task. Is it not very much more reasonable to suppose that the Creator, in designing man, should have made him an agent as well as an instrument, should have given him effective powers of teleological causation in order that he might co-operate in the Great Design; rather than to suppose that he endowed him with such mental powers only as would make him a helpless spectator of the drama, power to suffer without power to seek relief, power to desire and hope without power effectively to strive?

Henderson's demonstration does not in the least weaken

the case for teleological causation in organisms. Rather, in so far as it is conclusive, it greatly strengthens it by proving that Mind has operated effectively upon matter in the past and is therefore capable of doing so in the present. It removes, in fact, the chief bar to the acceptance of teleological causation in organisms, namely, the difficulty in believing that Mind can intervene effectively in mechanistic events.

Needham's third ground for preferring mechanistic to teleological explanations in science is, unlike the first and second, one that may have a certain degree of logical validity. 'Science can have nothing whatever to do with teleology, because it is not a notion which is susceptible of quantitative treatment. . . . Purposiveness is not a metrical concept. . . . Therefore, whatever we may think of teleology, we cannot have anything to do with it in the laboratory.' For 'anything that is in essence incapable of measurement is in essence incapable of scientific treatment'. From which it follows that 'whatever degree of realness teleology may have, it cannot be of any meaning in scientific investigation'. This is a restatement of the old dogma that 'science is measurement'.

The application of this dogma to the exclusion of teleological explanation in biology may be shown to be illegitimate in two ways. First, the dogma is of very questionable validity. In the greater part of biology and in much of geology, measurement of any but the roughest kind has hitherto been impossible; yet these branches of science have progressed and brought us much understanding and useful knowledge. How little does paleontology owe to measurement! Yet it is an important branch of science. Darwin's principle of natural selection owes little or nothing to measurement. Yet it has proved itself to be a most fruitful hypothesis and is commonly regarded as a scientific theory of the first importance. Henderson asserts 'the survival of the fittest has now become in the judgment of all biologists an unquestioned force in the moulding of life'. But by what methods and by what standards is this 'force' measurable? In short, biology has been built up on a vast mass of observations, very few of which were quantitative in any but the rudest sense.

But, if we accept this dogma that science may take notice only of measurable things and events, it does not rule out teleological events from the purview of science. Every branch of science has to make a beginning ; and in its early stages its measurements inevitably lack accuracy. Physical science itself did not begin with the establishment of exact standards and methods of measurement. All these were late products of an immense amount of research.

Biological science, deterred by the 'mental twist' in favour of mechanism, has only recently begun to attempt the measurement of the teleological factors in biological events. Yet it has begun and with good promise of progress. It is true at present that, as Needham remarks, 'no scale of intensity could be suggested. . . . for the striving of a blastula to grow into a chicken'. But that may cease to be true in the future. If we wish to apply measurement to teleological events we must begin, not with the microscopic blastula, but where the teleological character is most clearly in evidence, namely, with the actions of men and animals. Just now experimental psychology, which Needham in his lordly way pronounces to be the only scientific pschology, is much concerned with this problem. It is true that psychologists, with the fear of such autocrats as Needham strong upon them, generally claim to live and work within the fold of pure mechanism, and therefore use a terminology calculated to disguise the nature of their aims and purposes. Having grossly neglected the conative, the teleological, factors during the first fifty years of experimental psychology, they are now realizing that all this earlier work was very imperfect if not quite vitiated by this neglect. And they are now turning to the study in the laboratory of motivation, incentives, determining tendencies, governing propensities, and, above all, of 'drives'. That is to say, they are measuring the intensity of the conative factor, the strength of the impulse or impulsion towards goals ; their first efforts are not wholly unsuccessful and the methods will undoubtedly be improved by sustained co-operative effort.[1]

Henderson and Needham, in their discussions of teleology, make the common mistake. Instead of studying and reflecting upon actual teleological or purposive events of the most clearly marked type, they ignore these entirely ; and,

[1] While writing this note I received an article entitled 'An Experimental Analysis of the Obstruction Method of Measuring Animal Drives.' The authors (Drs. P. G. Warden and H. W. Nissen) conclude by saying : 'Our general conclusion must be that the obstruction method . . . fulfils most adequately the purpose for which it was designed.'—*Journal of Comparative Psychology*, Vol. VIII.

with some passing reference to 'the striving of a blastula', they concentrate their attention upon systems and organizations that seem to have been produced by purposive guidance of some sort. Contemplating the fitness of such systems for their functions under given circumstances (as Paley contemplated a time-piece) they regard such fitness as the only available evidence of teleology; whereas it is, at best, but an indirect and ambiguous evidence. Only by this obstinate blindness to the direct evidences of teleology are they enabled to say, as Needham says, 'there is as much teleology about the inorganic as about the organic'; to describe Henderson's argument as 'the final death of vitalism'; to relegate teleology to the remote creation of the material universe and to assert that 'Materialism . . . is the only philosophy upon which science can get to work'.

Like so many others, the authors of these two books say in effect: My purpose in writing this book is to show that purposive activities do not occur; the goal I seek is a final demonstration that goal-seeking is a process of a type with which we have no acquaintance, of which no instance is known to us. And the mere writing of the book is by their own confession a refutation of all the arguments they array.

The effort to demonstrate the unreality of purpose is, in short, a striking demonstration of purposive activity, a teleological activity that seeks to prove its own impossibility. In comparison with it the act of sawing off the branch on which one sits appears as a highly rational procedure.

By a happy coincidence, at the moment I had written the foregoing sentence and thus, as I thought, completed my manuscript, there came to me, through the kindness of its author, an advance copy of Mr. H. V. Knox's book, *The Will to be Free.* In this essay Knox shows that the attempt to establish universal determinism is an assertion of free will, and the acceptance of universal necessity an act of free choice. Knox's brilliant demonstration of the ineffective character of the argument for Determinism, his shattering attack upon the citadel of intellectualism, his withering exposure of the puerilities of conventional logic, embolden me to say a word on this topic. In view of the wide prevalence of the prejudice in favour of determinism, I had thought it wise to avoid raising this question. I will now venture to say that the reality of teleological causation, or purposive activity, does imply indeterminism of the will. As Knox puts it: 'the belief in the reality of purpose is identical with or necessarily implies—though for that very reason it cannot possibly prove—the existence of real possibilities in our world. For this belief, or faith, is manifestly the belief in the existence of a real field for the exercise of voluntary activity. It is thus at once the expression of the will-to-live and a presumption as to the nature of man's environment. It is, obviously, a necessary pre-supposition of practical action. But . . . it does not commit us to the view that possibilities have any pre-existence in "Nature" prior to their "discovery" by us. . . . Real possibility is, at one and the same time, the characteristic creation of the human spirit, and a function—in the strict scientific sense of that word—of real knowledge. For with increasing knowledge real possibility ever grows. And this means that neither pure necessity nor possibility is a fixed quantity in the world wherein we live and have our being. It is the special function of knowledge, throughout human history, to augment the second at the expense of the first. And that being so, it follows that the extent to which our intelligence actually controls the course of "Nature" exactly measures the extent to which our "knowledge" is real knowledge. Scientific experiment, in particular, rightly carries conviction, because what in every instance it really demonstrates is our ability to produce the given phenomenon *at will*. Every successful experiment in the scientific laboratory incontinently explodes the dogma that "Nature" is, in Whitehead's phrase, "closed to mind". And strange indeed it is that the sound of that explosion should have proved so inaudible to the ears of philosophers, and even to the ears of the "man of science" himself.

'Intelligence, as exercising control, is . . . indistinguishable from volition. The deterministic, which is the intellectualistic, treatment of will as merely an "object of intelligence" can now, therefore, be seen as a mere obscurantist device; devised, indeed, for the express purpose of maintaining the outrageous and otherwise indefensible fiction that real knowledge is essentially useless. . . . Will . . . is the point at which intelligence becomes itself an effective factor in the world which it "knows". It is a mode of intelligence; it is, in fact, intelligence in action. This voluntaristic conception of "will" as in principle identical with intelligence runs . . . directly counter to the intellectualistic, or deterministic, conception of " intelligence " as in principle " disinterested " ; or, in other words, as biologically inert and functionless. . . . For the biologist no less than for the philosopher, the final, comprehensive, and only consistent purpose of intelligence is to *transform* " Reality ", and mould it nearer to our heart's desire.'[1]

The two distinguished thinkers who write the preface to Mr. Knox's book concisely summarize its teaching: 'Determinism also is essentially a volitional adventure, vitalized by the very principle it attempts to discredit '. They add : ' That a universe ruled by necessity should give birth to the illusion of freedom and need the propaganda of determinism to set the matter right ; or that a universe of real options should give birth to a philosophy which denies their existence and need the counter propaganda of our author, remains, up to date, an unfathomed mystery.' But perhaps the mystery can be in some degree resolved. If the physical realm is wholly mechanistic and determined, and if Mind is teleological, capable of creating possibilities and of choosing between them, the present state of affairs is, after all, natural enough. Man's survival has depended primarily upon his efficiency in understanding and directing physical events. He has, therefore, been chiefly interested in them ; and this predominant interest has shaped the form of his language, the caste of his thinking, the structure of his mind. It has led him to develop the physical sciences in advance of and out of all proportion to the sciences of Mind. Hence

[1] Courage to declare that teleological causation is incompatible with the prejudice in favour of strict determinism comes to me from another source, namely, Professor A. S. Eddington's *The Nature of the Physical World*, which comes to hand at the same time as Knox's book. It appears that the developments of the quantum theory are leading the physicists to repudiate strict determinism in their world, and to put in its place, ' the Principle of Indeterminacy '. We learn that ' Classical physics foists a deterministic scheme on us by a trick ; it smuggles the unknown future into the present, trusting that we shall not press an inquiry as to whether it has become more knowable that way. . . . There can be no fully deterministic control of inorganic phenomena unless the determinism governs mind itself. Conversely if we wish to emancipate mind we must to some extent emancipate the material world also. There appears to be no longer any obstacle to this emancipation. . . . It seems that we must attribute to the mind power not only to decide the behaviour of atoms individually but to affect systematically large groups—in fact to tamper with the odds on atomic behaviour. . . . We must suppose that in the physical part of the brain immediately affected by a mental decision there is some kind of interdependence of behaviour of the atoms which is not present in inorganic matter.'

he feels confident and masterly in reasoning of pure mechanism; he falters and fumbles when he attempts to reason about himself. He finds a ready escape from his perplexity by setting up the fiction that he also is but a machine, and erects the fiction into a methodological principle of science. Yet, though the first effect of man's increasing control over nature has been to convince him that he has no control, it would be strange if, with further increase of that control, he should not reverse that first hasty conclusion.

It has been suggested by an English Bishop that we should call a halt to scientific research and devote ourselves for a period to adjusting human life to the results so far achieved. If this suggestion may be taken to mean that we should give primacy to the sciences of Life and Mind and somehow divert to them a large part of the human energy now directed to the exploration of the physical realm, it deserves enthusiastic support. For it is not only in the sphere of theory and speculation that our culture is rendered lopsided through the relatively excessive development of the physical sciences. In the sphere of practice, as is now increasingly recognized, our civilization is in danger of becoming self-destructive, just by reason of our lack of understanding of human nature and our consequent inability to shape and control our own development. The dogma that *Man is a Machine* is at once the perfect symbol of that lack of understanding and the greatest obstacle to the overcoming of that inability.

INDEX